OIL AND GAS PRODUCTION

Wireline Operations

by
Nancy Gore

Edited by
Jeanette Paxson

Published by
Petroleum Extension Service
Division of Continuing Education
The University of Texas at Austin
Austin, Texas

in cooperation with
American Petroleum Institute
Production Department
Dallas, Texas

1984

Published 1984. Second Impression 2005
Printed in the United States of America

Catalog No. 3.31010
ISBN 0-88698-043-7

Table of Contents

Foreword

Wireline Operations is one of a series of study lessons covering appropriate topics in oil and gas production. As with other lessons, this one is intended to contain information presented during lecture sessions of the API-sponsored School of Production Technology.

Wireline Operations covers the development of wireline operations, describes wireline equipment, and describes the various operations (diagnostic, troubleshooting, and completion and production maintenance) performed using wireline equipment.

The material presented in this book has been reviewed by a number of industry specialists including various members of the API Advisory Committee for the School of Production Technology, as well as instructors of that school. Striving to meet the educational needs of the petroleum industry, we welcome comments from the users of any of our training materials. In particular, we invite information concerning the relevancy of the material presented to current practices in the field.

Curtis Kruse, Director
Petroleum Extension Service

Acknowledgments

Sincere appreciation is extended to the many people who contributed to the preparation of *Wireline Operations:*

E. P. Lukert of Gearhart Industries, Bob Mouret and Ed Evans of N. L. McCullough, Jack Huffman of Otis Engineering Corporation, George Martin and Richard G. Ross of Camco, Harry Dieckman of J. C. Kinley Co., Barry Satterwhite of Welex Division, Halliburton Services, David Thompson of Broderick and Bascom Rope Co., Sally Allen of Baker Production Services, Brian Schailble of the Texas Railroad Commission, John Diller of Halliburton Services and Rick D. Berry of Homco International for acting as technical consultants and providing background information;

E. P. Lukert of Gearhart Industries and John Rushing of Otis Engineering Corporation for reviewing the finished manual;

Rob Trimble of the American Petroleum Institute, John Rushing of Otis Engineering Corporation, Rhonda Hooper of Totco, Malcolm Rosser of Halliburton Services, Glen Bliss of Camco, Rick D. Berry of Homco International, Mrs. J. C. Kinley and Harry Dieckman of J. C. Kinley Company, Colin M. McRae of Baker Production Services, Aileen J. Barr of Otis Engineering Corporation, and Ray McNiece of Welex Division, Hallibuton Services for providing photos and illustrations for the text.

How to Use This Manual

The format of this manual includes a set of specific objectives for each section; at the end of the section is a competency self-test. To get maximum benefit from the manual, read the specific objectives carefully before studying the material in each section. As you study the material in the section, take notes, using the objectives as a guide to the most important parts.

When you feel that you have mastered the objectives, begin the self-test. Since it is a self-test, *you* decide whether you should refer back to the material to answer the questions by determining how important that section is to your work. If you feel that you need to be very competent in an area, do not refer back until you have finished the test. This way, using the scoring points given at the beginning of the test, you can determine your percentage of competency. Score the test by using the corresponding key provided at the end of this manual.

1

Development of Wireline Operations

OBJECTIVES

Upon completion of this section, the student will be able to:

1. Define *wireline.*
2. Define *conductive wireline.*
3. List three types of nonconductive wireline.
4. Describe the advantages of small-diameter wireline.
5. Explain the difference between wireline and wire rope.

INTRODUCTION

Wireline is a strong, thin length of wire cable mounted on a powered reel. It may be conductive (electric) or nonconductive (mechanical) from 0.066 to 0.108 inch or from $^{3}/_{16}$ to $^{7}/_{16}$ inch in diameter, and as long as 30,000 feet. Wireline is used to raise and lower a tool or device in the wellbore. Procedures performed with tools that are suspended on a wireline are generally called *wireline operations.*

Wireline is particularly important during *completion,* the series of operations performed on a drilled, tested well to bring in the pay zone, and during *production,* the continuous well servicing and monitoring necessary to keep oil and gas flowing. Field operators can run anything from a basic downhole directional survey to the most delicate gamma ray formation log on wireline. They can fire perforating charges at precisely determined downhole locations, back off a string of stuck pipe, retrieve a wrench, or manipulate complex subsurface well pressure and flow controls.

Indeed, many of today's most vital oil field operations depend directly on the use of wireline. It is no wonder, then, that the line's usefulness was first discovered during early developments in one of the industry's basic procedures – oilwell cementing.

DEVELOPMENT OF NONCONDUCTIVE WIRELINE

Halliburton's Clothesline

In 1919, Erle P. Halliburton moved to the oil-rich area around Wilson, Oklahoma, to start a cementing business. He had been experimenting with the two-plug cementing system (fig. 1). Success with a two-plug job depended on the top plug's following the slurry to the bottom of the casing, where it would bump the bottom plug and stop the cement flow at its correct downhole location. A sudden increase in pumping pressure was supposed to signal a bump, that is, when the top plug bumped the bottom plug. It often meant, however, that the top plug had simply stuck in the casing. The only

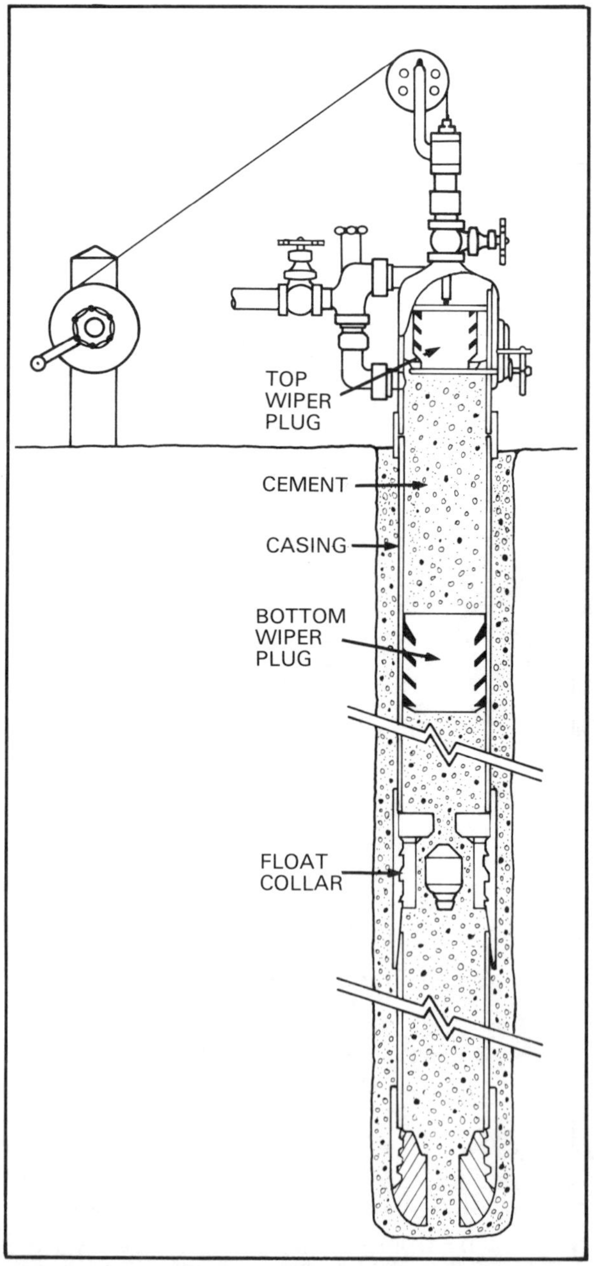

Figure 1. Two-plug cementing procedure using Halliburton's clothesline (Courtesy of Halliburton)

way an early cement crew could tell the difference was by guesswork; they estimated the pipe's fluid capacity and, when the top plug stopped, tried to determine whether the right amount of fluid had been pumped.

Like every major innovator in the history of the oil business, Halliburton decided there must be a better way to know what was happening downhole. He used a length of regular clothesline, marked it every few feet, attached it to a plain window-sash weight, and ran it downhole from a simple handcranked reel to measure the exact depth of the top cementing plug. Soon he substituted a numerical code for the marks on the line to keep a better record of the footage. There were drawbacks: the clothesline had to be cleaned and remarked after every job, and it tended to stretch. However, Halliburton had discovered a basic method of taking downhole measurements, and rapid refinements – both in technique and in the line itself – were inevitable.

Flat Line

Halliburton soon replaced his clothesline with a flat steel measuring line, which was marked at intervals like a surveyor's tape and which had a lead weight attached to its end. The flat steel line was both stronger and less apt to stretch than its predecessor, but its disadvantages soon rendered it equally unworkable. Improvements in rotary drilling techniques, for example, enabled operators in the early 1920s to drill far deeper than before. Obtaining flat tape of sufficient length – not to mention reeling it adequately – was a problem. Also, deeper wells contained higher pressures. Wireline measuring operations had to be run with a view toward containing that pressure, and flat line was difficult to pack off tightly in the pressure-containment stuffing box that had been designed to minimize pressure leakage around the line. Finally the weighted flat line stretched at great depths so that crews had to correct the readings to obtain an accurate measurement.

Round Steel Wireline

By the mid-twenties, flat measuring line had been replaced by circular, solid plow steel wire, which is still widely used today. *Plow steel* is a high-grade, extra-strong steel containing a high percentage of carbon. With a 0.066-inch diameter, this round wire can be wound in lengths of up to 10,000 feet. Its stretch is insignificant, and it can be packed off in a stuffing box with no problems.

At first, the round wire was marked at intervals like those on the clothesline and the flat wire used before it. The operator simply kept track of the amount of line entering the well. By 1926, however, Halliburton had invented the *well-measuring device* (fig. 2), which featured a calibrated

Figure 2. Standard measuring reel assembly with a measuring wheel similar to the one used on Halliburton's well-measuring device

measuring wheel that ticked off feet or metres as the line was spooled from the reel. The wheel proved more accurate and more convenient than the marked-line measuring method. The well-measuring device threaded the wireline around the wheel and through a stuffing box designed for pressure containment. The well-measuring device was, in a rudimentary form, a complete, self-contained system that met two of the fundamental requirements for a successful wireline operation: (1) *accurate* depth measurement and (2) pressure containment while running the line. As soon as these two requirements were met, the industry witnessed a staggering expansion of wireline capabilities.

Modern Nonconductive Wireline

Halliburton's refinement of the solid, nonconductive measuring line during the 1920s developed out of one basic problem: how to measure the progression of a top cementing plug down a well casing. By the early 1930s, operators and service people throughout the industry perceived that the line could have much wider application. Besides measuring from the surface to a downhole point of interest, nonconductive wireline could be used, for example, to install tubing plugs so that tubing could be run under pressure and to run subsurface recording devices to measure bottomhole pressure and temperature. These operations required heavier tool strings and thus stronger wire, and new types and sizes of nonconductive line were soon produced to perform the operations successfully.

Several types of nonconductive wireline have been produced to meet the requirements of the different types of wireline operations.

Slick line. Single-strand, nonconductive line with a 0.066- to 0.108-inch diameter is often called *slick line, measuring line,* or, more colloquially, *piano wire.* Slick line is available in lengths of up to 30,000 feet. Because the line carries heavy tools to great downhole depths, it is normally made of improved plow steel, which is chosen with a view toward low cost as well as high tensile strength.

Braided Line. When nonconductive (sometimes called *mechanical*) downhole jobs call for weight and force that would break slick line, wireline operators use stronger $3/16$- to $1/4$-inch braided, or stranded, line. Known in the field as *torpedo* or *well-shooter's line,* braided line usually consists of a thin fiber core surrounded by 5 × 7 (5 strands, 7 wires per strand) improved plow steel. This construction gives the line flexibility as well as very high tensile strength, and, like slick line, it can be used in lengths of up to 30,000 feet.

Corrosion-resistant lines. Hydrogen sulfide (H_2S), a gas often encountered in wells all over the world, can literally crystallize a plow-steel wireline. Operators may use a line made of stainless steel or of plow steel coated with nickel cadmium in an H_2S environment. Although both materials successfully resist hydrogen embrittlement and corrosion, they have disadvantages. Stainless steel is ultimately much weaker than improved plow steel and much more expensive. Nickel cadmium is prohibitively expensive for most operators. Alloys have been developed for use in deep, high-pressure wells with hostile environments (H_2S, CO_2, etc.). Some lines made of alloy cost as much as a dollar or more per foot.

DEVELOPMENT OF CONDUCTIVE WIRELINE

The development of a conductive electric wireline paralleled the refinement of solid, nonconductive measuring line. The need for increasingly sophisticated methods of evaluating downhole formations (logging) called for a line that could not only be measured for exact depth but also be used to relay electric signals to instruments at the surface. As early as 1926, Conrad and Marcel Schlumberger (pronounced "Slumber-jay") discovered that surface electrical measurements resulted in crude, hand-plotted logs indicating subsurface ore deposits. Further experimentation led them to develop, by the end of the 1920s, techniques using conductive line to take even more accurate subsurface measurements. By the end of World War II, the Schlumbergers had developed, in basic form, the conductive line used today for logging, perforating, and some types of pipe recovery.

Most electric wireline consists of a multi- or single-conductor copper core coated with an insulating sheath of Teflon and wrapped with two layers of braided plow-steel armor. Single-conductor lines were developed during a period of time in which Schlumberger held an exclusive patent on multiconductor lines. The two types simply have different transmission systems: whether a given wire is multi- or single-conductive affects neither its size nor its applicability. Electric wireline is generally spooled in lengths of 25,000 to 30,000 feet.

WIRELINE SIZES

Modern wireline is available in a wide range of sizes, as shown in table 1. Large-diameter wireline is often required for

TABLE 1
Sizes and Common Uses of Modern Wireline

Type of Wireline	Line Diameter (Inches)	Common Uses
Nonconductive	0.066	Measuring depth.
	0.072	Measuring inclination and pressure;
	0.082	Installing, retrieving, and manipulating downhole plugs, chokes, and other devices;
	0.092	Repairing and conditioning tubing; Removing light sand and paraffin; Light fishing and swabbing.
	0.108	Heavy fishing and swabbing; Removing heavy sand and paraffin. Used seldom and mostly overseas for jobs requiring line heavier than 0.092; 3/16- or 1/4-inch line is usually used instead.
	3/16	Heavy fishing and swabbing; Removing heavy sand and paraffin;
	1/4	Carrying heavy tool strings into deep wells for jobs requiring strong line.
Conductive	0.206	Running cased-hole logs; Jet perforating;
	3/16	Running free-point indicators to retrieve stuck pipe;
	7/32	Used in deep wells having pressures over 2,000 psi.
	5/16	Standard perforating, standard cased-hole logging, open-hole logging.
	7/16	

wireline operations because *tensile strength,* or resistance to rupture under tension, increases as line size increases. Whenever wireline operators have a choice, however, they use the smallest diameter of wireline suitable for a given job because small-diameter wireline has several advantages. With its low bend radius, a small-diameter line can be wound around a small reel drum, which is both more convenient and less expensive to transport than its larger counterpart.

Another advantage of small-diameter line is its light weight, which makes transportation and line handling easier. In addition, particularly during deep downhole operations involving heavy tool strings, heavy wireline ($^5/_{16}$-inch line weighs 180 pounds per 1,000 feet, for example) increases the braking requirements of the surface engine powering the reel.

Finally, a small-diameter wire has a small cross-sectional area and requires less weight applied at the surface to move it through pressure-containment devices and to lower it downhole. At 10,000 pounds per square inch (psi) well pressure, for example, a 0.066-inch line requires only 34 pounds of weight to pull it through the stuffing box, whereas a $^5/_{16}$-inch line requires 767 pounds.

Figure 3. Wire rope used to raise and lower the traveling block on a drilling rig.

WIRELINE AND WIRE ROPE

Wireline and wire rope are both long, flexible lengths of wire, usually made of steel. Both are used in the oil field to lift and lower tools and equipment for basic operations. There the similarity ends, although industry-wide confusion between these two terms is great and persistent. The confusion is understandable. Field personnel have long used the terms interchangeably. Publications sometimes label as *wireline* the heavy rotary drilling lines in and around a rig, or as *wire rope* the torpedo line used for downhole swabbing and fishing. The consequences of this confusion are somewhat less than serious, of course, but a few definitions can nonetheless unravel the issue.

Wire rope is a large-diameter braided line used strictly for heavy mechanical operations. A common type of wire rope is the rotary drilling line employed on almost all drilling rigs (fig 3). Ranging in length from 2,500 to 10,000 feet and from 1 to 1¾ inches in diameter, wire rope (called *drilling line*) is used on the rig to raise the traveling block, hoist equipment up the derrick, and perform other surface operations involving heavy lifting and pulling.

Also categorized as a type of wire rope is *sand line,* which is used to perform downhole operations such as sand control, coring, and swabbing when the job is too heavy for the heaviest (⅜-inch) braided wireline. The most common diameter of sand line is $9/16$ inch, but it can range from ⅜ to ¾ inch. It consists of a solid fiber core surrounded by 6 × 7 (6 strands, 7 wires per strand) braided steel.

Wireline and wire rope can be distinguished, then, according to both function and size. As opposed to the much heavier drilling line, wireline is used exclusively for subsurface, never surface, operations. While some overlap exists in the use of torpedo line and sand line, the former has a smaller 5 × 7 construction and a maximum diameter of ⅜ inch. Finally, wireline service companies carry neither drilling nor sand line on their units.

WIRELINE SERVICES

As wireline began to be applied extensively throughout the gamut of oil field operations and as methods for operating the line grew more sophisticated and complex, companies began to specialize specifically in wireline work. Today, a wireline service company is almost always called in to perform a wireline operation. An exception has to do with basic depth and direction measurement during drilling: a large drilling company occasionally installs its own small measuring unit near a rig or just above the rig drawworks. Not much bigger than a desk, this unit consists of a small motor and reel hookup, a length of slick line, a simple depth counter, and a deflection survey instrument designed to measure the degree of deviation (fig. 4).

Figure 4. Small wireline-measuring unit for installation on a rig (Courtesy of Halliburton)

Although some operators have their own units that they use for bottomhole pressure surveys, wireline companies (or wireline divisions within large service companies) perform nearly all work requiring the use of nonconductive or electric line. These wireline companies or divisions are often specialized according to specific types of operations offered, and each has a variety of servicing capabilities.

Figure 5. Wireline mounted on a truck

On land, wireline units are usually mounted on trucks that are backed up to the well (fig. 5). Another type of land unit is a van with a crane mounted on its roof for hoisting the lubricator on top of the well (fig. 6). Marine units are mounted on skids and transferred to the site on boats or barges (fig. 7). In terms of the size of its reel and the amount of horsepower it is carrying, a given unit may be light- or heavy-duty. The type of unit depends on the complexity of the job to be performed.

Regardless of its size or the job it is required to do, however, each unit contains a reel of wireline; a power system to play out and retrieve the line; instrumentation indicating weight, line speed, and depth of the tool string; and the specific tools necessary for the operation (fig. 8). Lubricators ensuring adequate pressure containment are also probably included.

Figure 6. A van equipped with a crane for hoisting the lubricator on top of the well (Courtesy of Amoco)

Figure 7. A skid-mounted wireline unit for use offshore

Figure 8. Interior of a wireline unit containing instrumentation and specific tools necessary for a wireline operation (Courtesy of Amoco)

Self-Test

1: Development of Wireline Operations

(Multiply each correct answer by five to arrive at your percentage of competency.)

Fill in the Blank

1. ______________________ is a strong, thin length of wire cable mounted on a powered reel.

2. Procedures performed using tools that are suspended on a wireline are generally called ______________________ ______________________.

3. Wireline was first used while performing a ______________________ operation on an oilwell.

4. The first wireline used was a ______________________.

5. Wireline that can relay electric signals to instruments at the surface is called ______________________ wireline.

6. List three types of nonconductive wireline used today.
   (1) ______________________
   (2) ______________________
   (3) ______________________

7. ______________________ ______________________ or resistance to rupture under tension, becomes greater as the line size is increased.

8. List three advantages of using the smallest possible diameter of wireline.
   (1) ______________________
   ____
   ____
   (2) ______________________
   ____
   ____

(3) ____

____

____

True or False

9. ______ The terms *wireline* and *wire rope* are interchangeable.

10. ______ Wire rope is a large-diameter braided line used strictly for heavy mechanical operations.

11. ______ Sand line is used to lift equipment on the surface that is too heavy for regular wire rope.

12. ______ Wireline and wire rope can be distinguished because wireline is used exclusively for subsurface operations while wire rope is used on the surface.

13. ______ Well completion is the series of operations performed on a drilled, tested well to bring in the pay zone.

14. ______ Tensile strength increases as line size decreases.

15. ______ The two fundamental requirements for a successful wireline operation are (1) accurate depth measurement and (2) a way to run the line under pressure.

2

Wireline Equipment

OBJECTIVES

Upon completion of this section, the student will be able to:

1. Describe the considerations that must be taken into account when planning a wireline operation.
2. List the three types of reel and power systems.
3. List the three types of measurement indicators used with a wireline tool string.
4. List the parts of a nonconductive wireline tool string.
5. List the parts of a conductive wireline tool string.
6. Describe the precautions that should be taken to prevent damage to a wireline.

INITIAL CONSIDERATIONS

The equipment carried by a typical wireline unit enables its operators to perform a wide array of oilfield operations. Nonconductive line operations, for example, are used to perform anything from basic downhole measurements to complex fishing and sand control jobs. An electric line unit may be used to run a logging survey, to perforate casing during completion, or to locate the exact position of a length of stuck pipe.

The specific tools and procedures used for a given operation depend, of course, on the exact nature of the job at hand. Therefore, planning the operation—choosing the appropriate line size, making sure required downhole tools and equipment are available, and establishing pressure control requirements involves a number of considerations. The wireline operator needs some idea of the depth to which the line is to be run, as well as of downhole pressure and temperature. If the requested job involves cased-hole logging or perforating, for example, the operator needs to know the casing diameter and weight. These factors are especially important when tools with close tolerance are used. Field personnel requesting a wireline service should be able to answer these questions and any others that may pertain to the needed work.

Despite the specific requirements of individual operations, all units share basic requirements necessary for any wireline procedure. A wireline job is accomplished with a reel of slick or conductor line, a power system enabling operators to raise and lower the line and to control its speed, instruments indicating weight and depth of the downhole tools, and equipment allowing for operation under pressure. A tool string provides a way to attach tools to the line.

REEL AND POWER SYSTEMS

The earliest wireline unit, which was a small reel wound with a short length of wire, was operated manually. As wells rapidly grew deeper and as wireline began to be applied more and more extensively in all phases of oil and gas production, muscle power became inadequate. Heavy upward jarring operations called for far more speed than cranking by hand could provide. The sheer weight of the tool string as it was lowered as far as 10,000 feet could not possibly have been controlled without mechanical assistance.

The Sidewinder

Oilmen borrowed a source of power from the automobile industry, and by the 1930s, they were able to operate wireline under these more demanding conditions. Power was transferred from the engine of a truck to the reel of wire in what was nicknamed the *sidewinder unit.*

The sidewinder (fig. 9) was used by most wireline companies until the middle 1950s. It is still used by some small operators today. On a sidewinder, a spool of wire is pulled from the left side of the truck. A belt is connected to the truck's transmission system by way of a pulley on the left rear wheel, and the belt is tightened over the pulley. To achieve pull, or speed, the left rear wheel is jacked off the ground. Then the worker in the truck puts the transmission in gear and operates the gas pedal to obtain the necessary power. Several methods of remotely controlling the accelerator are also used.

The sidewinder's advantage is, of course, that it allows the same wide range of power and speed control as the engine of a truck does. Its disadvantage is that, without remote accelerator control, it requires two

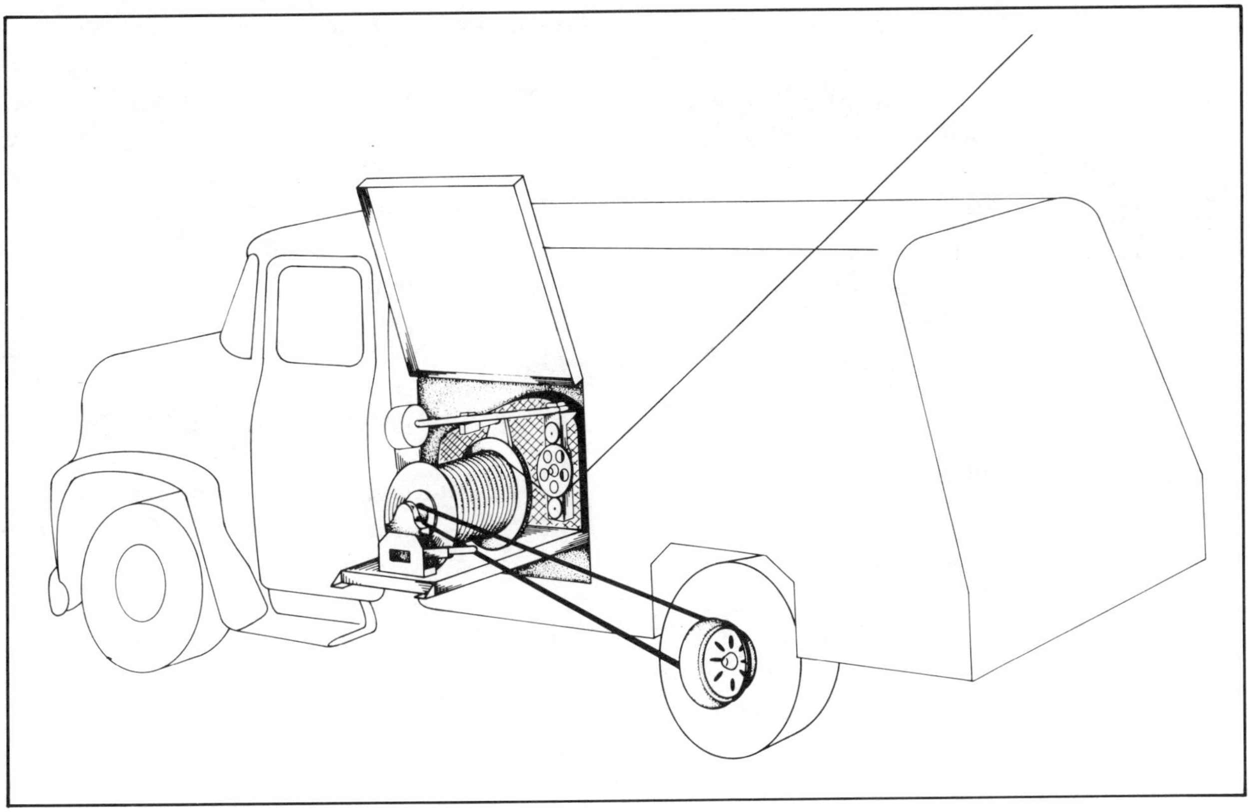

Figure 9. A sidewinder

operators—the driver and an outside operator—to keep track of the line and its progress. Close communication is required whenever a change in line speed is needed, and directions from outside often are not carried out fast enough, for example, to keep a rapidly rising tool string from crashing into the stuffing box and breaking the line.

The Hydraulic Improvement

Around 1948, when wells had become both far deeper (up to 23,000 feet) and more crooked than they had been before, a hydraulic system was developed to power the reel of wireline. A system using an electric generator and a chain to drive the reel was also invented. Both of these systems are commonly used today. Hydraulic power, however, remains the method of choice for the great majority of today's wireline operators for a number of reasons.

First, this system, which takes diesel or gasoline power from the engine of the truck and transfers it to a hydraulic motor, allows one operator to monitor the line and to control power simultaneously. Too, a bypass valve between the pump powering the reel and the reel motor allows the operator control over both the rate of the line's descent and the amount of upward pull. This valve can be operated like the throttle on a car to disengage power from the line, and the line can be completely stopped with one touch of the control lever. Most hydraulic units are fitted with three- or four-speed transmissions to afford tremendous speed and power ranges. Depending on the size of an individual unit, power can range from 10 to over 100 horsepower.

The Dual Reel

The reel on most modern wireline units is about 4 feet in diameter. It sits in the center on the back of the unit for easy accessibility as the unit backs up to the well. Many operators desire the versatility of more than one size of wire on the unit at the same time, so often a unit carries two reels. For example, one reel of slick line and one of torpedo line allows the operator to perform a difficult, high-pressure job with the smaller wireline and then perform swabbing with the larger line.

INSTRUMENTATION

Nowhere, perhaps, is the sophistication of today's wireline unit more apparent than in its instrumentation panel. The operator can monitor at a glance an enormous number of conditions pertaining to the line and its progress into and out of the hole (fig. 10). The conditions that require monitoring during a specific operation vary according to the job. Instruments are available for wireline units in a wide variety of combinations suitable for almost all wireline operations. However, some devices are more important, and thus more common, than others. The most important device, and one appearing on every unit, is an instrument used to monitor the line's downhole depth.

Depth Indicators

Regardless of the specific wireline job to be performed, the operator must be aware of the depth of the tool string at any given moment. During perforation, for example, shaped charges must be fired at a downhole location that is exactly adjacent to the producing zone. Also, during any operation, the operator must be aware of the location of end-string tools or instruments as they ascend to the wellhead in order to control speed and to prevent the tool string from running into the stuffing box. The measuring system invented by Erle P. Halliburton in the 1920s is still commonly used today

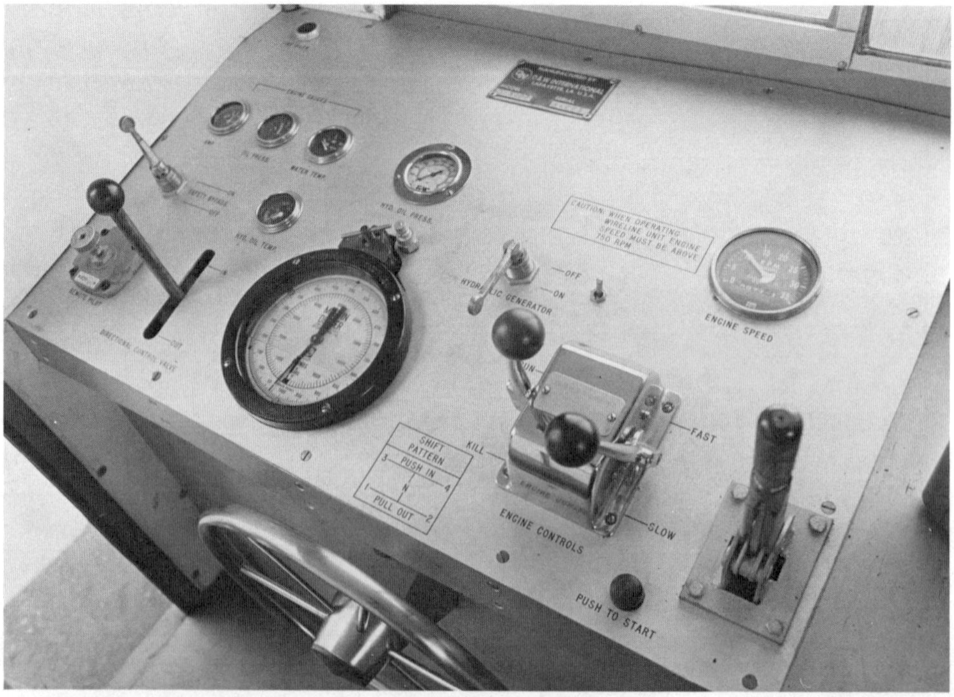

Figure 10. Wireline instrumentation panel

(fig. 11). On its way into and out of the hole, the line travels around a measuring wheel that drives a counter or odometer calibrated in feet or metres.

On most conductive line units, the line travels around a small-diameter measuring sheave in front of the reel. The sheave's revolutions drive a small motor that in turn indicates the measured depth of the tool string. A second odometer, which can be reset at any time and then run simultaneously with its counterpart, allows the operator of a perforating job, for example, to correlate the wireline depth with depths indicated on pertinent logs.

Weight Indicators

Each size and type of slick line and electric line has a breaking point at which a tool string that is too heavy or that has been run too deep causes the line to fray or split. Table 2 indicates the breaking points of some commonly used line. Ordinarily, an operator maintains the weight on the string at a level far below the allowable one. In

TABLE 2
Breaking Points of
Some Common Wireline Types and Sizes

Type of Wireline	Size of Wireline (Inches)	Breaking Point (Pounds)
Slick Line	0.066	810
	0.082	1,240
	0.092	1,550
Conductive Line	3/16	3,600
	7/32	5,200
	5/16	10,300
	7/16	17,900

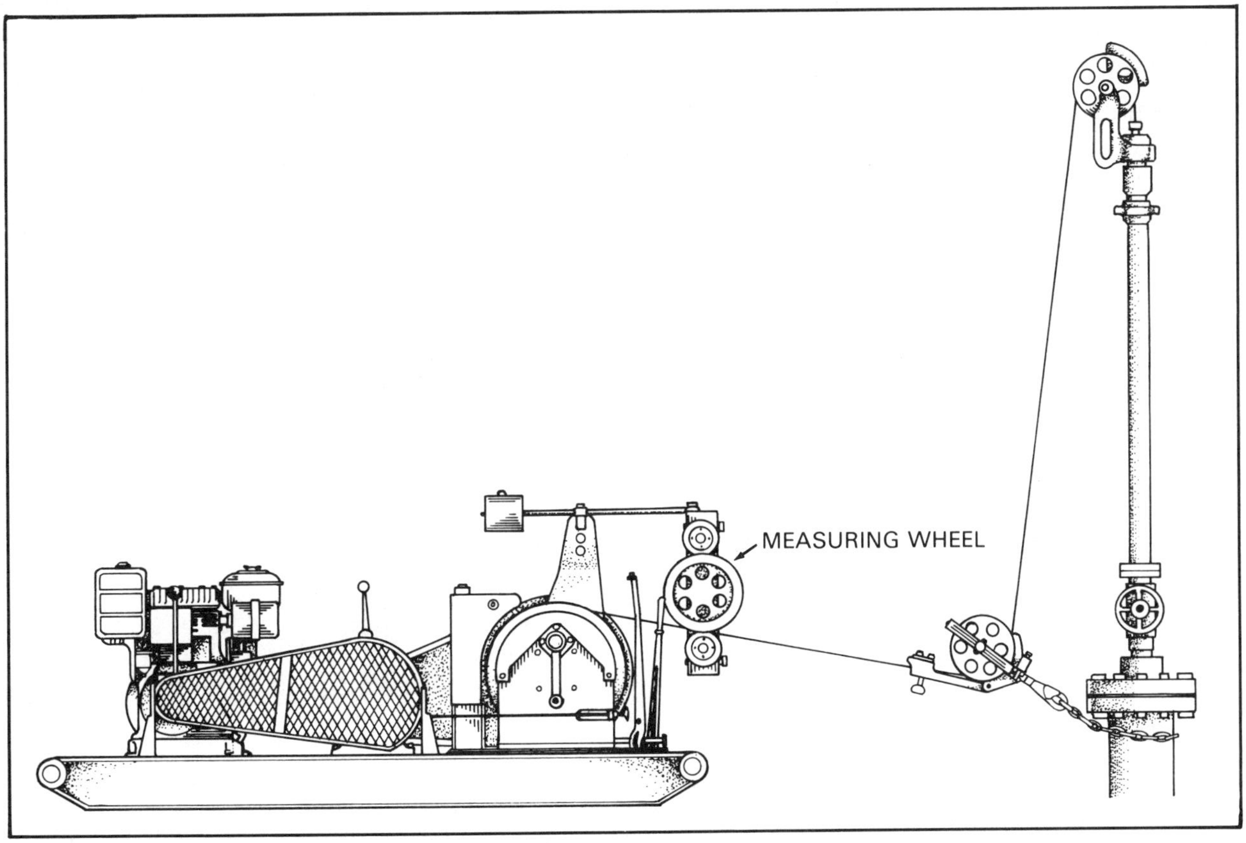

Figure 11. Measuring system in which the wireline travels over a measuring wheel that drives a counter or odometer calibrated in feet or metres (Courtesy of Halliburton)

most cases, a simple tug on the line indicates to the operator whether the weight is within bounds. However, in particularly heavy wireline work, such as heavy swabbing with large mechanical jars or pulling on the line when the tools become stuck, the line must be weighted to its maximum safe load. In these instances, some type of weight-indicating device must be used.

A weight indicator may be mechanical, hydraulic, or electronic in design. It is calibrated in pounds to indicate the total load or line tension. It may be attached to the depth-measuring device or built separately into the control panel.

Speed Indicators

A delicate logging or temperature survey calls for a downhole instrument to be raised at a precisely determined, constant speed. A *line speed indicator* is an instrument calibrated to show this speed in feet per minute. The indicator may be driven mechanically by the measuring wheel and placed next to the depth odometer, or it may be driven and recorded by a motor.

THE TOOL STRING

The *wireline tool string* is the string of tools and equipment attached to the end of the slick or conductive line and lowered into the hole. The diameter of the wireline tool string may vary from ¾ to 1⅞ inches depending on the diameter of the tubing or casing through which the line is to be run. A tool string 1½ inches in diameter is the most popular size. Length, chosen according to desired weight and the deviation of a given hole, varies from 1 foot to 7 or 8 feet. Tool strings and lubricators for logging and perforating jobs may be 20 feet or more in length.

Nonconductive

Beginning at its point of attachment to the line, the tool string used in most mechanical (nonconductive) operations consists of a wireline socket, a stem, jars, a knuckle joint, and the tool needed for the job (fig. 12).

The *wireline socket,* sometimes called the *rope socket,* connects the wireline to the tool string. The wireline socket contains a rotating disk, which allows for horizontal swivel between the line and the tools below it. The socket also contains a small spring that cushions the line and prevents it from breaking during heavy operations.

The *stem* provides the weight required to perform a jarring operation or to overcome high pressure. For that reason, the stem is also known as the *weight bar* or *sinker bar.* A fishing neck on the upper end of the stem is used when the tool string becomes stuck in the tubing. The line is cut or pulled from the socket and removed from the borehole. A fishing tool is lowered and attached to the fishing neck on the stem in order to retrieve the string.

The *jar* has been called the workhorse of a wireline job. It is designed to be quickly extended or collapsed lengthwise. Surface manipulation of the line is used to deliver rapid upward or downward blows to equipment that is being retrieved from or set in the tubing. The *stroke jar* is a mechanical jar that can deliver either upward or downward impact. An operator may add a hydraulic jar for a particularly strong upward blow or an explosive jar for an especially heavy downward impact. These jars should never be substituted for mechanical stroke jars.

The *knuckle joint* contains a ball-in-socket arrangement to allow angular movement between the jar and the end-string tool. This

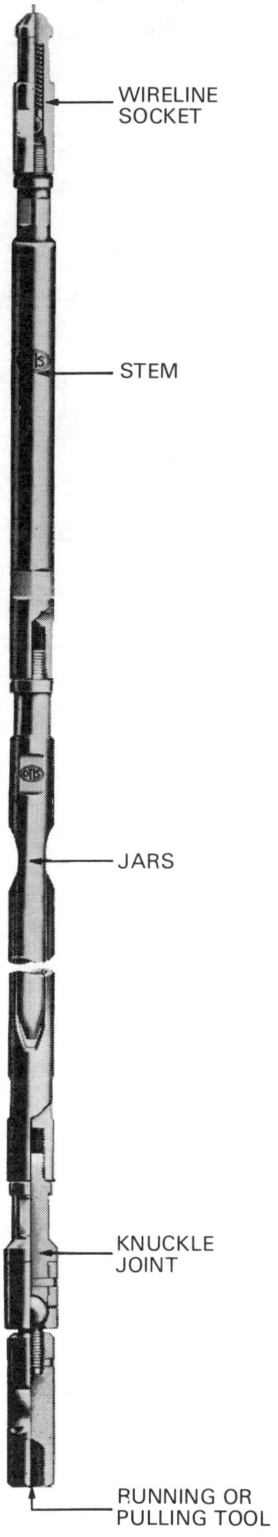

Figure 12. Mechanical wireline tool string (Courtesy of Otis Engineering)

flexibility is especially important in crooked tubing (most tubing is crooked, to some extent). When an exceptionally long tool string is used, or when the tubing is especially crooked, a knuckle joint may be placed not only in its usual location between the jar and the end-string tool, but also between the stem and the jar.

The *end-string tool,* often simply called a *wireline tool,* is the lowermost component of the tool string. Depending on the job, this device may be any of a large number of end-string tools.

Conductive

Wireline operations requiring the use of electric line (logging, perforating, and free-point detection) generally do not require the weight and force necessary for most mechanical jobs. A typical tool string for conductive line may consist of (1) a *fishing neck* often called a *head,* which is used so that the line can be pulled from it if the tool becomes stuck and so that a fishing tool can be latched onto it for retrieval; (2) a *turn buckle,* which connects the fishing neck to the end-string tool and allows flexibility of the tool string within the tubing or casing; and (3) the specific logging, perforating, or free-point detection devices to be used for the job.

PRESSURE CONTAINMENT

The wireline tool string is set up at the surface, but it must perform downhole, where pressures may be in excess of 30,000 psi. To bring about this transfer from the surface to the downhole environment, while simultaneously containing well pressure, is a complex procedure involving a number of considerations.

Blowout Preventer

When wellhead pressures are high or when a well contains dangerous hydrogen sulfide, a wireline blowout preventer (fig. 13) must be attached to the wellhead before other pressure containment equipment can be installed. The wireline blowout preventer, also called the *wireline valve,* is recommended as a safety precaution on any job, regardless of a well's known or expected pressure. The wireline blowout preventer is a valve designed to close around the wireline and to shut in the well if high pressure threatens a blowout. The blowout preventer is a wireline crew's most important piece of pressure control equipment.

Installation of Pressure Containment Equipment

The wellhead for a flowing well is often called the Christmas tree, because of its many valves, spools, gauges, and horizontal attachments. To perform an operation, the wireline crew typically closes off the uppermost valve on the central vertical series of valves and the wing valve, which allow the well fluids to flow into the flowlines to the left (fig. 14). Pressure above these valves is bled into the atmosphere, and the top section of the tree is removed.

The *gin pole,* a set of three 8-foot interconnected tubular poles telescoped on site to a length of approximately 20 feet, is first attached to the well with a *chain and load binder* (fig. 15). The gin pole is used to set the wireline blowout preventer or wireline valve onto the top of the wellhead.

Then, the lubricator and stuffing box are assembled near the wellhead. They are threaded with the wireline and its tool string and are hoisted on top of the wireline valve. The *lubricator* is sometimes referred to as the *riser assembly.* It is usually a three-part section of tubing rated to withstand high pressures. The diameter and total length of the lubricator are sufficient to hold the entire tool string (including any tool or

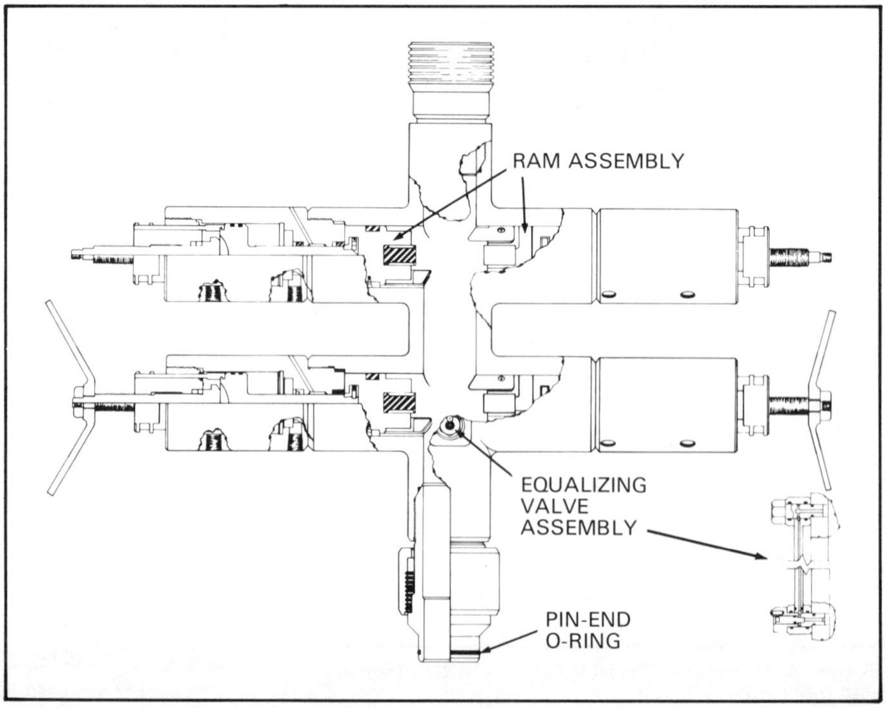

Figure 13. A wireline blowout preventer (Courtesy of Otis Engineering)

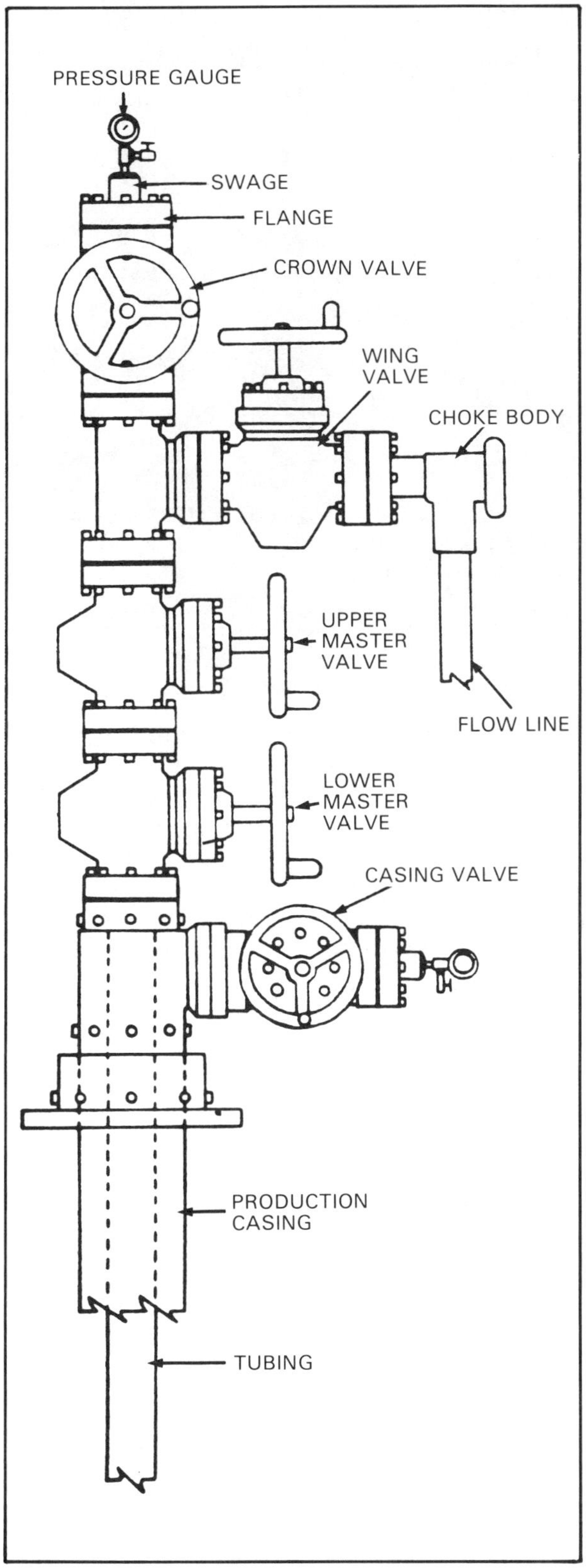

Figure 14. Typical wellhead (Courtesy of Otis Engineering)

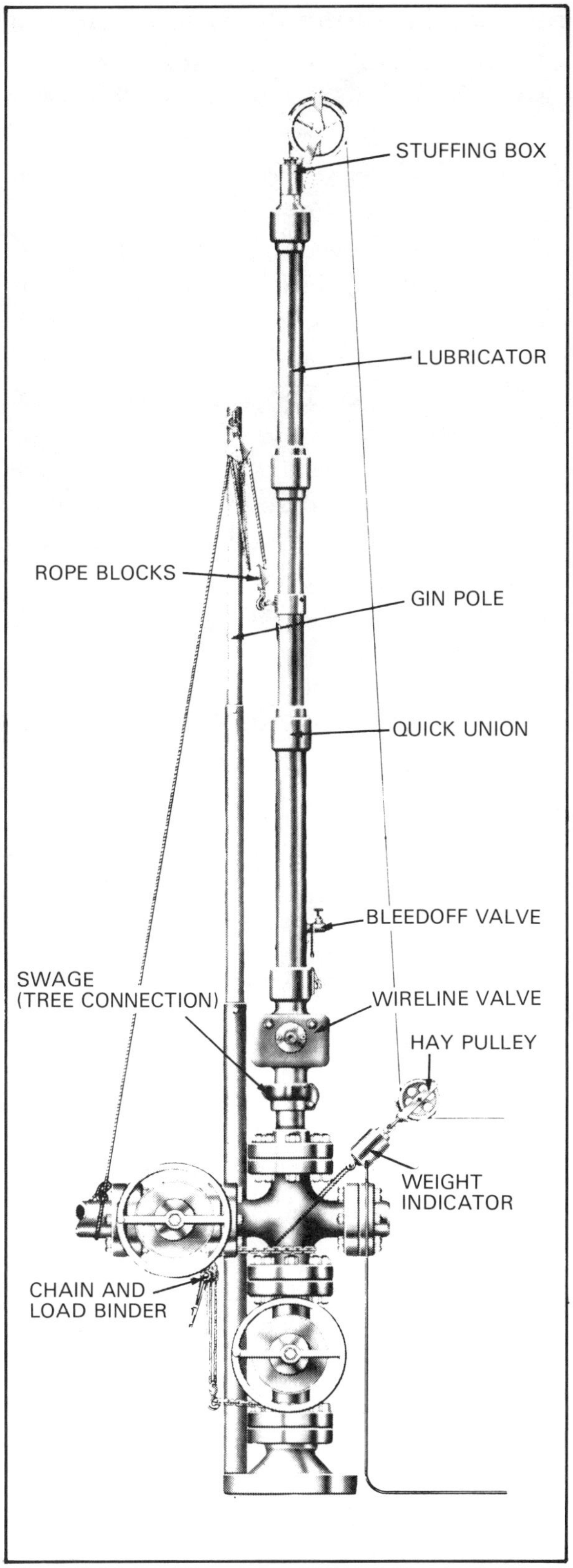

Figure 15. Wireline surface equipment

device to be run in the well or removed from it) between the stuffing box and the swab valve. Well pressure admitted into the lubricator may be vented through bleeder valves, which may also be used for installation of pressure gauges.

The *stuffing box,* which sits on top of the lubricator, consists of a packing chamber containing packing materials used to form a tight seal in order to keep well pressure leakage around the wireline to a minimum. A *top sheave* attached to the stuffing box is used to guide the line as it travels from the reel into the box and down into the well. If the line breaks and is forced out of the packing, a small blowout preventer built into the stuffing box automatically seals off pressure. A tool trap is sometimes located near the bottom of the lubricator in order to prevent accidental loss of tools if the wire is pulled out of the rope socket at the top of the lubricator.

After the lubricator and stuffing box have been fully assembled on the ground (the wireline threaded through the stuffing box and attached to the tool string, which has been placed in the lubricator), the entire assembly is raised, set on the wellhead, and securely attached to form a pressure-tight seal above the wireline valve. Then the line is run under the *hay pulley,* and any slack is spooled onto the reel. The upper master valve on the wellhead may now be opened for movement of the wireline tool string down the wellbore.

When the job has been completed, or when the tool string needs to be removed or changed, the tool string is raised back into the lubricator and the upper master valve is closed. Any pressure within the lubricator is then slowly bled into the atmosphere, and the lubricator-stuffing box assembly is taken off the wellhead.

CARE AND HANDLING OF WIRELINE

Despite some of the rough-and-tumble downhole operations for which wireline is used, the line itself is rather fragile. It breaks easily, for example, when excessive weight is placed on it. If it is made of plow steel, it may rust if well fluids are not removed. Uncontrolled slack may cause the line to kink. Operators can avoid the loss in money and time that results from damaged wireline by following a few basic preventive steps.

The line should be washed as soon as it is brought up out of the wellbore. Otherwise, mud, oil, and water may spray the well site and may create safety and housekeeping problems as well as line rust. Washing may be done by hand, or the line may pass through a *line wiper,* a small tubular device sometimes connected to the hay pulley (fig. 16). The line wiper contains a rubber plug that encircles the line and strips it of all fluid. After the line has been cleaned, it should be oiled before it is stored.

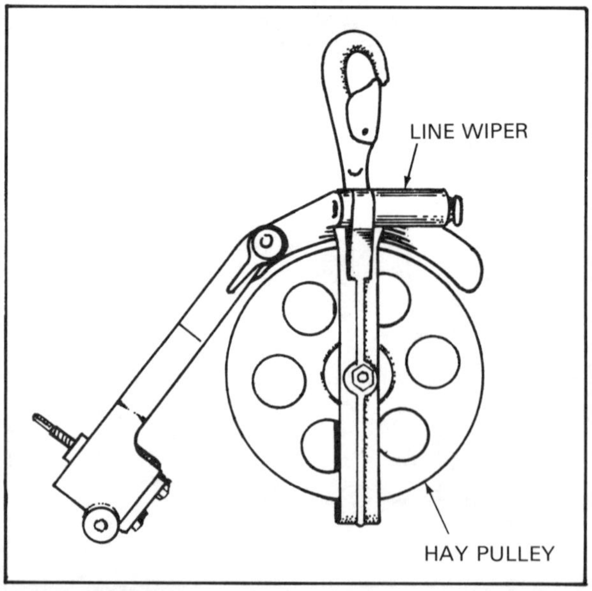

Figure 16. Hay pulley with a line wiper (Courtesy of Halliburton)

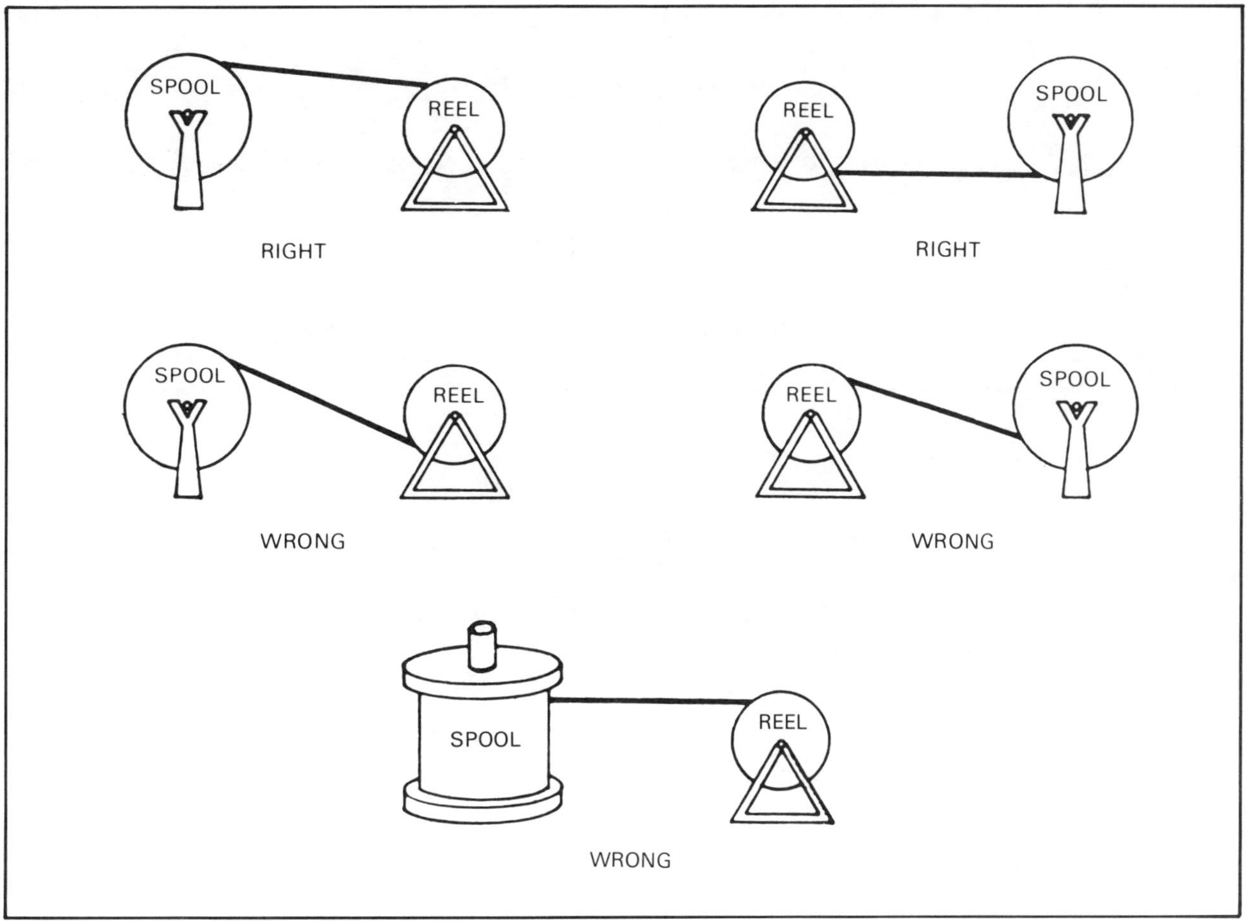

Figure 17. Recommended methods of transferring wireline from spool to reel (Courtesy of API)

Damage to the wireline may also be avoided by transferring the line correctly from the spool to the reel (fig. 17). Otherwise, twisting and kinking may cause the line to fray or break.

Finally, the amount of weight placed on a wireline should be kept as far below the line's breaking point as possible. Many wireline service people operate under the principle that the fragile line should be treated like a soft-shell egg in this respect. If the breaking point is 1,100 pounds, for example, a service company may recommend that the operator try to pull 880 pounds, or 80 percent of the breaking strength of the line, or fewer to establish a margin of safety and to a avoid costly, time-consuming line fishing job. The line should be designed to pull out of the fishing neck before tension on the line causes it to break.

Self-Test

2: Wireline Equipment

(Multiply each correct answer by two to arrive at your percentage of competency.)

Fill in the Blank

1. List the factors that must be considered when planning a wireline operation.

   (1) ____________________

   (2) ____________________

   (3) ____________________

   (4) ____________________

2. List the basic requirements that are necessary for any wireline operation.

   (1) ____________________

   ____

   (2) ____________________

   ____

   (3) ____________________

   ____

   (4) ____________________

   ____

   (5) ____________________

   ____

3. The source of power for the earliest wireline unit was ____________ operation.

4. List one advantage and one disadvantage of the sidewinder.

   Advantage: ____________________

   ____

   Disadvantage: ____________________

   ____

5. List the two types of power reels that were developed after the sidewinder.

   (1) ____________________

(2) ______________________________

6. List the three types of measurement indicators on the instrument panel of a wireline unit.

(1) ______________________________

(2) ______________________________

(3) ______________________________

7. List the components of a nonconductive wireline tool string.

(1) ______________________________

(2) ______________________________

(3) ______________________________

(4) ______________________________

(5) ______________________________

(6) ______________________________

8. List the components of a conductive wireline tool string.

(1) ______________________________

(2) ______________________________

(3) ______________________________

9. A wireline blowout preventer must be used when

(1) ______________________________

(2) ______________________________

10. The ______________ ______________ is the most important piece of pressure-control equipment.

11. Label the parts of the illustration below.

12. List four basic procedures to prevent damage to wireline.

(1) ______________________________

____

(2) ______________________________

____

(3) ____

____

(4) ____

____

True or False

13. ______ The sidewinder type of reel is still used today by some small operators.

14. ______ Manually operated reels became impractical as taller derricks were used for drilling and as wells were drilled to greater depths.

15. ______ In a nonconductive wireline tool string, the stem connects the wireline to the tool string.

16. ______ If a job involves cased-hole logging or perforating, the operator needs to know the diameter and the weight of the casing.

17. ______ The gin pole is used to set the wireline blowout preventer onto the top of the wellhead.

18. ______ A line wiper is someone who washes off the wireline as it comes out of the hole.

19. ______ The three parts of the conductive wireline tool string are the fishing neck, the head, and the turn buckle.

20. ______ A weight indicator may be mechanical, hydraulic, or electronic in design.

21. ______ It is important in any wireline operation for the operator to know the exact depth of the tool string.

3

Diagnostic Operations

OBJECTIVES

Upon completion of this section, the student will be able to:

1. List the reasons why it is important to drill a well as straight down into the earth as possible.
2. Describe the most widely used method of controlling hole deviation.
3. Explain why well logging provides some of the most important diagnostic functions available to the petroleum industry.
4. Describe the purpose of sidewall sampling.
5. Describe the two types of caliper surveys.
6. Describe the procedure for checking for sand fill at the bottom of a string of tubing.
7. Explain why temperature and pressure are important in drilling operations.

DETERMINING DOWNHOLE CONDITIONS

When wells were drilled only a few hundred feet into the ground, a blind guess was just as effective as any other method of determining downhole conditions. However, random guesswork worked only haphazardly and for short periods of time; as well depths increased and as oil field procedures became more sophisticated, the need for exact and reliable methods of determining downhole conditions increased.

By the 1920s, Erle P. Halliburton had discovered that wireline could be used to perform one of the oil field's most fundamental diagnostic procedures–measuring from the surface to a point of interest downhole. Marcel and Conrad Schlumberger had proved that electric line could be lowered into a well to make a simple, hand-plotted map of formation characteristics. The application of wireline as a diagnostic tool has grown so rapidly since these early innovations that today there are over fifty formation logging tools, for example, and more than ten separate wireline methods of taking wellbore temperatures. It has been estimated that throughout the life of an individual well, wireline will be lowered downhole at least 300 separate times for diagnostic purposes.

CONTROLLED VERTICAL DRILLING

A routine rotary drilling operation normally requires that the hole be drilled as straight down into the earth as possible. There are a number of very good reasons for limiting deviation, not the least of which is that state regulatory agencies require the driller to obtain special permission to drill directionally. In Texas, for example, a well owner must submit proof to the Texas Railroad Commission that the degree of deviation between bottomhole and surface locations corresponds to strict requirements set forth in the original well permit. If the bottomhole location is not within the required parameters, the well owner must shut in the well and apply for a spacing exception. Some additional benefits of controlled vertical drilling are that it maximizes the useful life of each drill bit, that it creates fewer problems with workover and artificial lift, and that casing can be set with more ease and speed in a straight hole.

In very deep holes, directional surveys may be run by dropping a recording instrument down the drill pipe during a connection and retrieving it with sand line. Drill pipe measurements are used for depth determinations. The most widely used method of controlling hole deviation, however, is to run a recording instrument through the drill pipe on standard measuring-sized slick wireline after every few hundred feet have been drilled. The *deviation and azimuth indicator,* which is also called a *double recorder* (fig. 18), contains an angle indicator, a timing element, and a replaceable chart. The timing element, which is set to go off when the tool has come to rest just above the drill bit, descends on the sharply pointed angle indicator, which then punches a hole in the chart. The chart rotates 180 degrees and a second hole is punched (thus the name *double recorder*) to verify a correct reading of the resting angle of the tool. The recorder is encased in a sturdy steel barrel called a *go-devil* (fig. 19), which is designed to protect the sensitive tool from shock, pressure, fluid, and dirt. Radial flexible spear points on top of the go-devil center the entire assembly within the pipe.

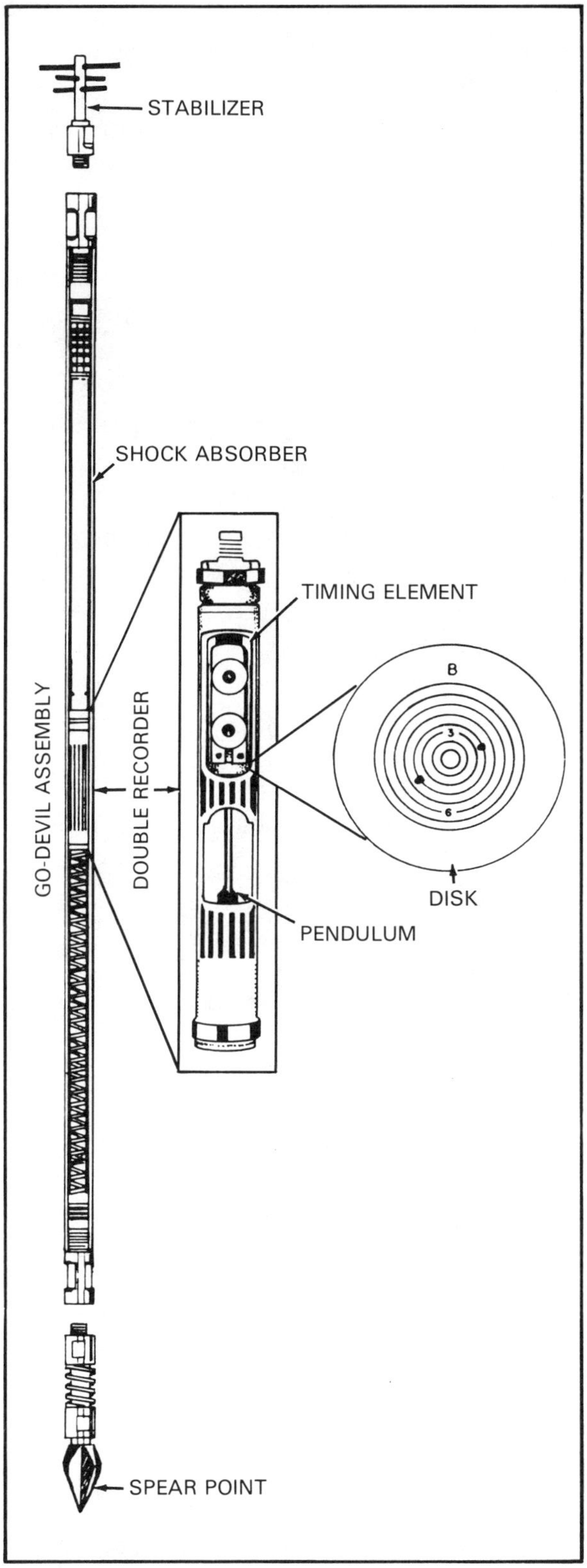

Figure 18. Double recorder, disk, and go-devil assembly equipped with stabilizer and spear point

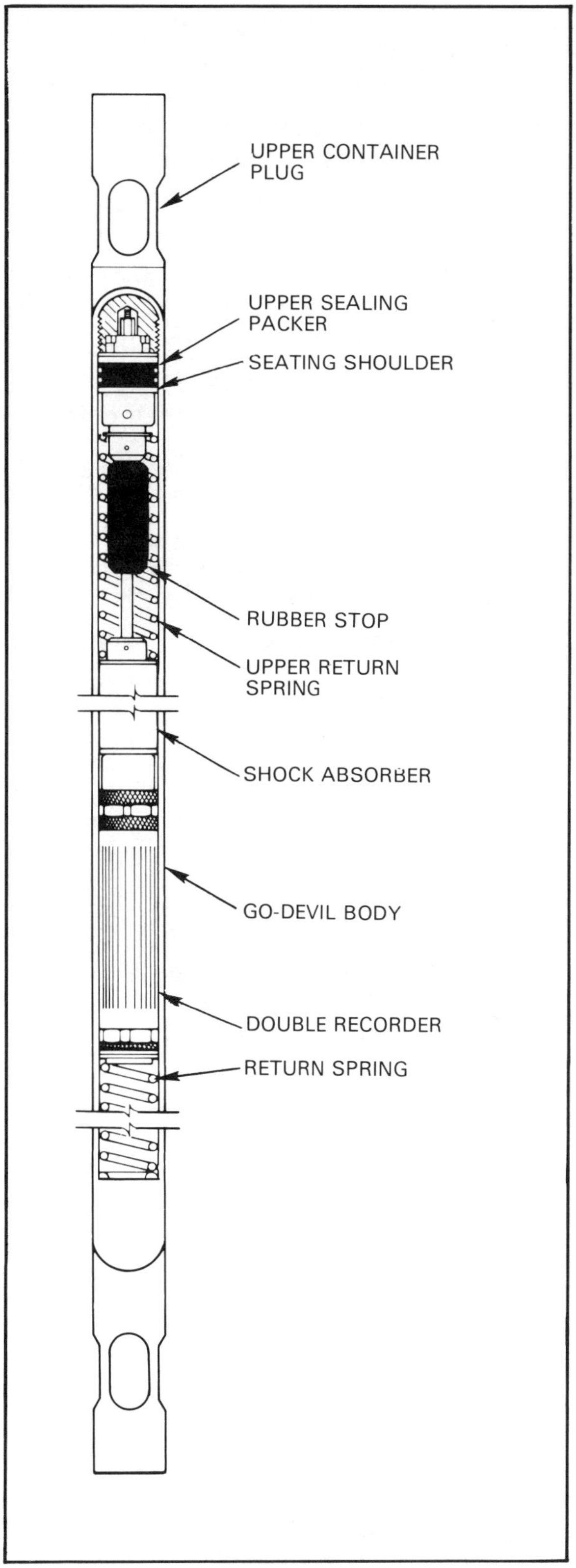

Figure 19. Go-devil (Courtesy of Totco)

Since a directional survey must be run frequently during drilling in order to keep the drill string on a vertical course, it is most often performed by the driller, who uses a small measuring unit that is installed near the rig or just above the drawworks. The measuring unit is equipped with a small motor and reel hookup, a length of slick line, a simple depth counter, and the recorder/go-devil assembly.

FORMATION LOGGING

Well logging is performed exclusively by crews from electric wireline service companies. These companies provide some of the most important diagnostic functions available to the petroleum industry. For example, an open-hole logging survey of a newly drilled well can reveal the presence of formation hydrocarbons and can predict whether the hydrocarbons will flow. Thus, an open-hole logging survey is a critical source of information for determining whether to complete the well and is, understandably, performed in practically all of today's drilling operations. Similarly, cased-hole surveys, run after casing is set and throughout the life of the well, are used to evaluate the effectiveness of the cementing operation, to monitor changes in wellbore hydrocarbon levels, and to assist in recompletion of older wells.

To log a well – whether open-hole or cased-hole – the wireline crew lowers one or

Figure 20. Inserting the radioactive source into a logging sonde

more long, cylindrical instruments called *sondes* into the wellbore. A logging sonde (fig. 20) is a highly sophisticated electronic instrument designed to measure and record electrical, radioactive, or acoustic properties of formations and to transmit the signals up the line while the sonde is being raised at a predetermined rate of speed. In the control cab of the wireline unit, these signals are translated – increasingly, via computers – into graphs that can be interpreted by trained geologists and engineers. The graphs are commonly recorded onto photographic film and many logging units contain a small darkroom where the film can be developed and printed.

Typically, both the open-hole and cased-hole crew consists of two operators and an engineer. The operators are responsible for making up the tool string and for lowering it into and raising it out of the hole. The engineer, who has usually been trained extensively by the service company to operate sophisticated computers and to evaluate logs (fig. 21), is responsible for overseeing the operation and making sure it goes smoothly.

While open-hole and cased-hole logging crews share certain basic functions and characteristics, they also have important differences. Indeed, both of these types of logging have become so sophisticated over

Figure 21. Evaluating logs using computers in the logging truck

the past forty years that the larger logging companies have split up into completely separate open-hole and cased-hole service divisions. Even small logging companies tend to specialize exclusively in one of the two service groups.

Open-Hole Logging

A well owner usually calls an open-hole unit to the wellsite during the final tripping-out of the drill string. Thus, the derrick is usually completely in place when the wireline operation is performed. Typically, the sonde is assembled on the catwalk (fig. 22) and hoisted to the floor via the air hoist. The logging crew then assembles a special sheave above the rotary, and the wireline from the reel on the truck is reeved through the sheave and attached to the sonde. The sonde is then lowered into the hole.

If the hole is full of drilling mud – and it usually is – it is not necessary to install pressure-control equipment before running the log.

Some open-hole logging operations take about 24 hours to perform. A number of logs may be run, depending on the requirements of an individual well, but several are almost always run. The *self-potential* (SP) curve is run to determine lithology, and a *caliper log* is run to determine the diameter of the wellbore and to

Figure 22. Assembling the logging sonde

locate any corrosion, scaling, or pitting problems inside the tubing. The *gamma ray log* measures and records radioactive waves caused by the existence of radioactive elements–uranium, potassium, and thorium–in some shales. Since these elements have very low permeability, a high gamma ray curve indicates low production possibilities. Two types of *resistivity logs*–also called *electric logs*–measure a formation and its entrapped fluids' resistance to electricity next to the borehole and more than two inches away. High resistivity correlates with low saltwater saturation and with a higher probability that porous spaces are filled with hydrocarbons. Finally, a *porosity log* is run to measure the density of a formation.

Cased-Hole Logging

Cased-hole diagnostic services are performed from the time casing is set until the well has ceased to produce. Occasionally, a workover rig may have just preceded a cased-hole crew to a wellsite and may still be rigged up. A *workover rig,* which is a kind of scaled-down drilling rig used to pull tubing from the wellbore and to perform various workover operations (fig. 23), has equipment for lifting and hoisting the logging equipment that the wireline crew uses. Usually, however, the cased-hole tools must be installed with heavy equipment carried on the wireline unit or on a separate truck. This equipment may include a mast for hoisting. It is, however, smaller than the workover rig.

Electric logging surveys cannot be run in a cased hole because metal casing distorts the electric currents. Other logs performed on an open hole, however, can be used in a cased hole. Usually these logs are run in conjunction with the collar log or gamma ray log for depth correlation for comparison

Figure 23. Workover rig

purposes. By checking formation logs after a well has been producing, an operator can gain information to assist in determining the rate of hydrocarbon depletion.

While the use of open-hole logging is restricted to evaluating the potential productivity of a new well, cased-hole logs have much wider application. They are used to perform pressure and temperature surveys and to determine where to perforate well casing and where to set completion tools.

SIDEWALL CORING

Sidewall coring, as its name implies, is a method used to cut *cores,* or samples of formation, from the wall of an open borehole and to bring them to the surface for analysis. Typically performed after drilling and before completion, sidewall coring is commonly used to supplement other tests (logging, mud sampling, drill stem testing) to evaluate whether a given rock formation will produce hydrocarbons at an acceptable rate. A site geologist may request a sidewall core when mud samples have been too contaminated with the drilling fluid to be truly representative or when an examination of a piece of rock may be needed to see how oil or gas are distributed through the formation. In any event, sidewall coring may be used to evaluate formation porosity, permeability, and fluid saturations, although often sidewall cores are too small or are too broken up by the impact of the bullet to provide reliable data. Sidewall coring can also provide data concerning formation minerals and evidence of fossils.

Since the most widely used wireline coring procedure, often called *sidewall sampling,* does not require that the drill string be lowered downhole again, it is far cheaper and less time-consuming than a conventional coring job that uses a core bit on the end of the drill string. Too, while wireline-obtained cores are small (about 1¾ inches long and 13/16 inch in diameter), they may be analyzed to determine porosity, permeability, and saturation. A trained geologist can quickly examine them for evidence of formation fluids or fossils. Thus, they provide physical evidence of data gathered by formation logging, and they can also help in determining whether more detailed testing should be run on the well before completion.

Sidewall sampling is performed using conductive wireline and a tool called, predictably enough, a *sidewall sampler* (fig. 24). This four-feet long, gun-like tool is honeycombed along its length with hollow cylindrical bullets attached to cables. When

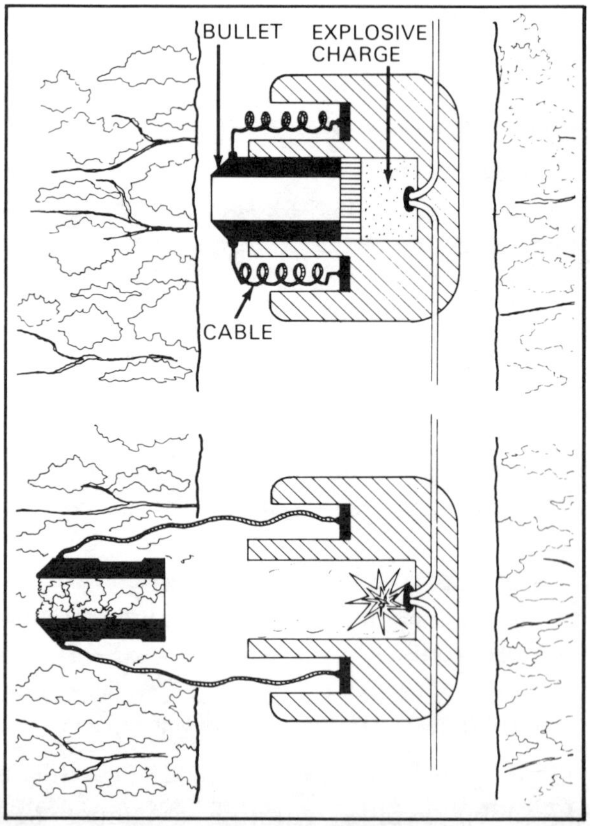

Figure 24. Sidewall coring gun operation

the wireline operator has determined, through standard wireline depth-measuring devices, that the lowered gun's charges are precisely adjacent to zones selected for coring, the bullets are fired. Shot either simultaneously or one at a time, the hollow bullets penetrate the formation, bite off pieces of rock, and retain them. After all the bullets have been fired, the entire assembly is raised to the surface, and the cores are delivered for analysis. By combining guns, the wireline crew can take up to 72 sidewall cores during one downhole trip.

CHECKING TUBING OR CASING

As formation hydrocarbons are depleted, the well's rate of production gradually decreases. This decrease in the rate of production is an unavoidable reality of the oil field and a normal occurrence that is not only expected but planned for by field operators. Sometimes, however, a flow rate shifts erratically, drops off dramatically, or stops altogether. The cause of such a problem is often a distortion or obstruction in the tubing string or casing: sand can enter the wellbore and block the upward flow of oil or gas; a tubing hole, caused by corrosion or pitting, can deflect the hydrocarbons out of the tubing and into the tubing-casing annulus; or a length of pipe can cave in, or collapse, because of extreme force from the adjacent formation. When production slowdown indicates one of these downhole conditions, a wireline crew is often called in to determine the exact nature and position of the problem.

Running a Caliper Survey

One common method of diagnosing a problem involves running a caliper tool (fig. 25) into the wellbore on either electric

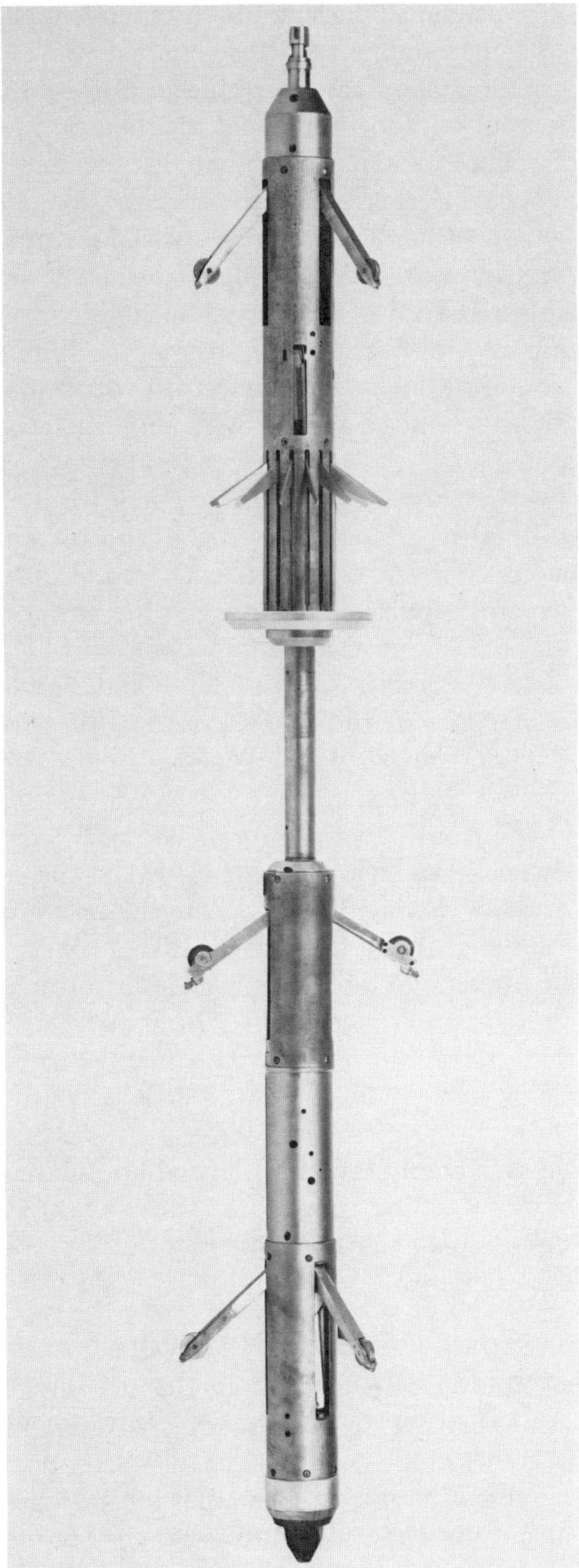

Figure 25. Microscopic caliper for 7-inch pipe (Courtesy of J.C. Kinley)

or mechanical wireline. Both types of tools, the mechanical and the inductive, are approximately two feet long and from 1½ inches to 3½ inches in diameter. One mechanical type contains two or more spikes, or arms, attached vertically along its length. When the caliper has been lowered to the bottom of the well, the arms are released with a spring inside the tool and are allowed to touch the sides of the tubing or casing throughout the return trip. As an arm reaches any irregularity in the pipe – such as a cave-in, sand buildup, heavy corrosion, or a tubing hole – it will retract or extend according to the dimensions of the irregularity. Each caliper arm generates pulse frequencies to the surface, and these are recorded on a type of logging chart. After assessing the depth and dimension of the obstruction or distortion, the production crew can then decide on how to correct the problem.

The caliper tool run on mechanical wireline is designed to pick up internal corrosion or flaws (pitting, erosion, wear) in the inside diameter (ID) of the pipe. This tool consists of fifteen or more feelers arranged around the circumference of the 52-inch cylinder. Within the cylinder is a chart that is turned by a drive wheel. As the tool is raised through the pipe, each feeler marks a line on the chart, revealing any abnormalities and providing operators with a three-dimensional cross-sectional view of the pipe's interior.

Tagging Bottom for Sand Fills

Often a producing zone produces so much sand that the area above the perforations becomes clogged and halts or slows production considerably. A common procedure for diagnosing this problem is to tag the bottom of the hole to check for sand fill. To perform this mechanical wireline operation, the wireline crew attaches a tubing end locator and a sample bailer to the end of the tool string (fig. 26) and lowers the string into the well.

The *tubing end locator* contains a spring-loaded finger, called a *dog,* that rotates when the tool enters an area of increased diameter – which is usually at the bottom of

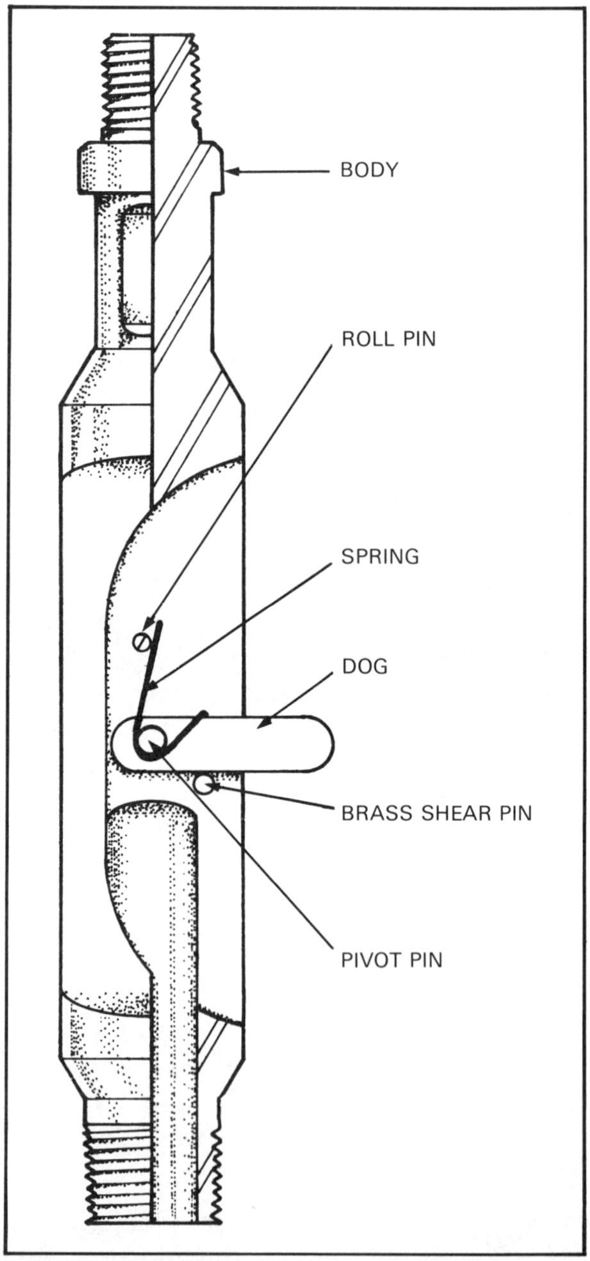

Figure 26. Tubing end locator and sample bailer (Courtesy of Camco)

the tubing but which may be higher if a tubing break has occurred. The dog then pops out horizontally to restrict movement back up into the smaller tubing area. The *sample bailer* has a sharp pointed end designed to slice into sand and a hollow cylinder to trap sand samples. After the dog has popped out upon reaching the tubing's end, the wireline operator runs the tools several times between that point and the bottom of the hole, setting the bailer on the bottom with enough impact to force sand into it. The tubing end locator is then jarred upward with enough impact to shear the brass pin and allow the dog to rotate downward and into a slot in the body of the tubing end locator. Then the entire assembly is again raised to the surface. If the operation diagnoses bottomhole sand fill as a problem, the production crew can evaluate a number of options for cleaning out this annoying obstruction.

Finding Holes in Tubing

Tubing holes, cracks, and breaks allow oil and gas to escape into the annulus and can slow production significantly on an artificial lift well. Although it causes little loss of production in a flowing well that has a packer, a tubing leak may still cause excessive pressure on the casing and may expose the casing to possible corrosion. The tubing end locator, and often the caliper tool, can be used to check for the size and location of this type of problem. Another way of diagnosing tubing holes involves attaching a length of soft line, or fiber rope (called a *pony tail*), to the end of the mechanical tool string. The assembly is then lowered into the hole. During the downward journey, fluid is pumped down the tubing from the surface, forcing the fiber rope into any hole that may be encountered. The tool string is thus stopped adjacent to the top problem spot.

RUNNING TEMPERATURE AND PRESSURE SURVEYS

Determining bottomhole temperature and pressure, as well as determining temperature and pressure variations along the length of the wellbore at any given time, can serve many important purposes. One reason to check the bottomhole temperature is that most downhole tools are not equipped to handle temperatures over 400°F. When temperatures exceed that mark, special tools must be used. Temperature surveys are often employed soon after cementing operations to locate the top of the cement column, because the setting of cement causes a measurable temperature increase. Surveys can also be used to determine the point downhole at which fluid is entering or exiting the wellbore. This survey helps to determine whether perforations are open and whether there are leaks in the casing string. The most common reason for running a pressure survey is to check the downhole pressure periodically to evaluate reservoir performance. A pressure test can also detect channels behind pipe and determine whether a high-pressure zone above a low-pressure zone may be deflecting gas downward and impeding its upward flow.

One method of running a basic bottomhole temperature test involves lowering a mechanically operated temperature-recording tool into the well with slick wireline. The tool is equipped at its lower end with a maximum-reading thermometer that registers the temperature after a few minutes in the hole. After the wireline crew has let this tool sit downhole long enough for the temperature reading to climb to its true level, the tool string is raised to the surface, and the temperature on the thermometer is noted.

To check temperature anomalies along the length of the tubing, an electric line operator may run a basic conductive-line temperature tool (fig. 27). This tool uses a very sensitive temperature-sensing element to detect the most minute changes in borehole temperatures, and it relays its findings

Figure 27. A basic conductive-line temperature tool

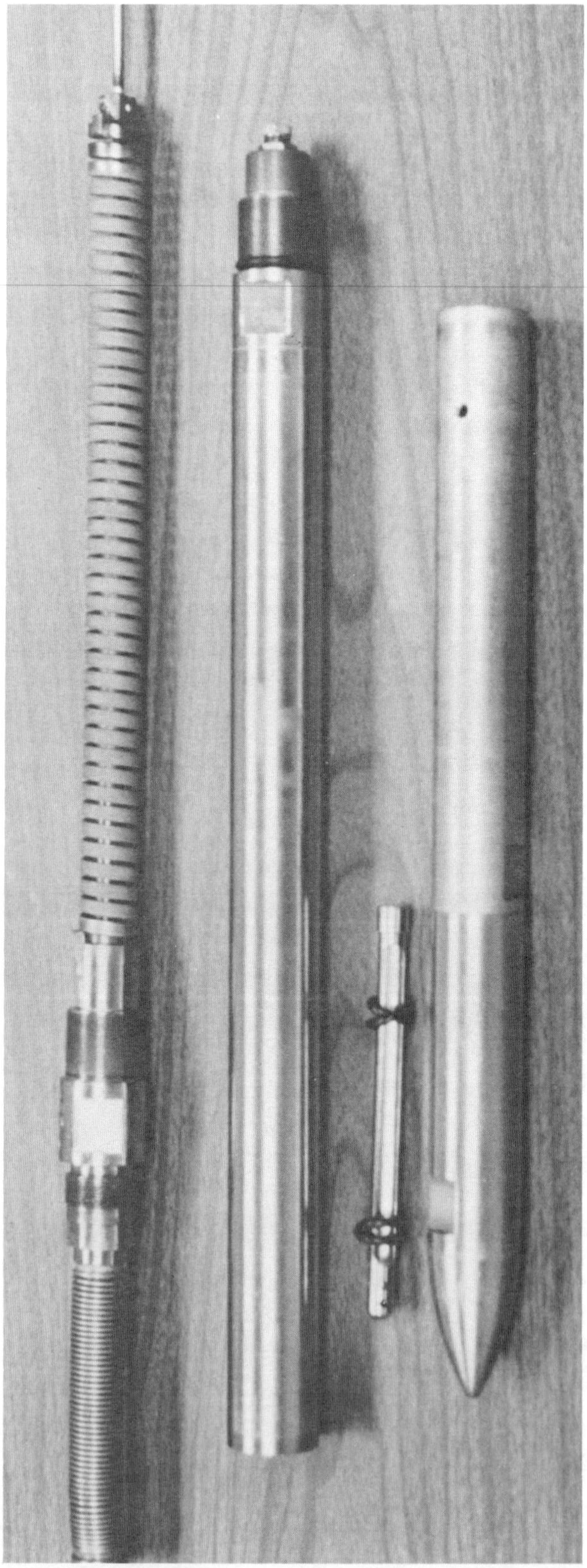

Figure 28. Amerada bomb used to take temperature and pressure surveys

up the line immediately. The result is an accurate curve displaying changes in temperature along the tubing string.

A common method of checking wellbore pressure and temperature involves running a bottomhole pressure gauge on mechanical wireline. Developed by the Amerada Oil Company in the 1920s, this gauge is still referred to by many today as an *Amerada bomb* or *Amerada gauge* (fig. 28). It has three basic parts: a clock, a recording section, and either a pressure element or a temperature gauge, depending on the data needed. If both pressure and temperature information are needed, the temperature gauge with its recording section and the pressure gauge with its recording section may be run in tandem on the same tool string.

One of the components in the pressure element, the Bourdon tube, rotates as pressure changes occur. This movement is transmitted directly to the recording section. This section has a scribe with a small metal arm with a needle, or stylus, that etches marks directly on a coated metal chart. The chart's position beneath the scribe is controlled by the clock. The scribe rotates as the pressure or temperature changes, and over a period of time, etches a curve on the chart.

Self-Test

3: Diagnostic Operations

(Multiply each correct answer by four to arrive at your percentage of competency.)

Fill in the Blank

1. Some benefits of controlled vertical drilling are that it maximizes the useful life of each ______________ ______________, it creates fewer problems with ______________ and ______________ ______________, and that ______________ can be set with more ease and speed in a straight hole.

2. The most widely used method of controlling hole deviation is
____
____
____

3. List the three parts of a double recorder.
(1) ______________________________
(2) ______________________________
(3) ______________________________

4. The long, cylindrical instrument used to log a well is called a ______________.

5. List the three types of properties of formations that are measured by a sonde.
(1) ______________________________
(2) ______________________________
(3) ______________________________

6. List five types of logs run in an open hole.
(1) ______________________________
(2) ______________________________
(3) ______________________________
(4) ______________________________
(5) ______________________________

7. Sidewall coring is typically used to supplement what three other tests?

(1) ______________________________

(2) ______________________________

(3) ______________________________

8. List three possible causes of a decrease in production through tubing.

(1) ______________________________

____

(2) ______________________________

____

(3) ______________________________

____

9. A ____________ ____________ ____________ and a ____________ ____________ are used to tag the bottom of a hole to check for sand fill.

4

Troubleshooting Operations

OBJECTIVES

Upon completion of this section, the student will be able to:

1. Describe the most commonly needed wireline troubleshooting operations.
2. List three methods of freeing stuck pipe.
3. Describe three tools used to repair and condition tubing.
4. Describe the two types of tools used for sand bailing.
5. Describe the tools most commonly used for fishing.

CORRECTING DOWNHOLE PROBLEMS

The need for troubleshooting results, as its name implies, when a problem downhole is so severe that it requires remedial measures. Production has probably dropped due to a collapsed area of pipe or due to sand buildup inside the tubing. Perhaps a leak in the tubing is allowing gas to escape from the tubing string into the tubing-casing annulus, or perhaps a piece of junk downhole is blocking the vital passageway between the perforated zone and the surface.

In any event, both mechanical and electric wireline units offer an array of troubleshooting procedures and tools to correct these and other problems. Some of the most commonly requested wireline troubleshooting operations include freeing a length of tubing, casing, or drill pipe just above the point where it is stuck downhole so that the stuck portion can then be retrieved; repairing or conditioning tubing; removing sand or paraffin deposits from strategic downhole locations; and fishing for dropped tools or various pieces of junk downhole.

FREEING STUCK PIPE

Because sand cave-ins, well washouts, doglegs, or junk in the hole occasionally cause a length of tubing, casing, or drill pipe to become stuck downhole, effective methods of freeing the pipe above the point at which it is stuck have been developed. Using electric line, these methods involve locating the point at which the pipe is stuck and then separating the pipe just above that location. Methods of removing the free section of pipe from the stuck section include string-shot backoff, jet cutoff, and chemical cutoff.

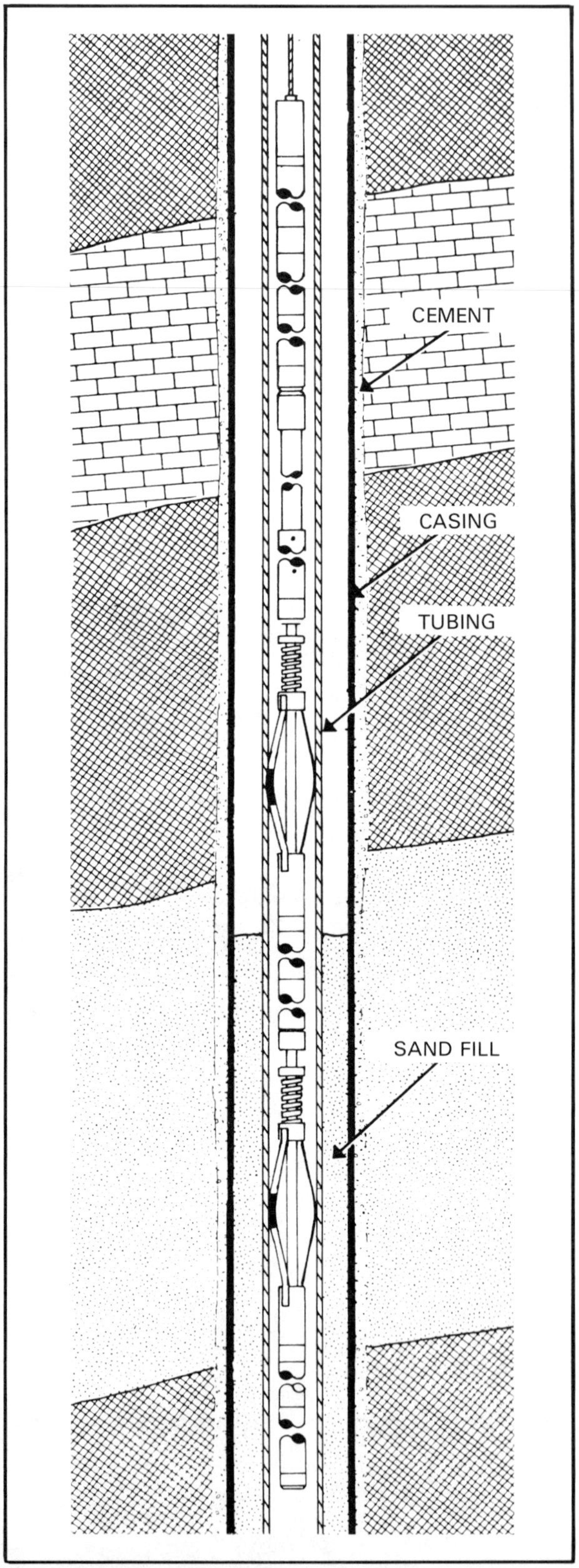

Figure 29. Free-point indicator

Free-Point Indication

A *free-point indicator* (fig. 29) is run downhole on electric line to determine where pipe is stuck. As the tool is run, electrical impulses from two contacts against the inner wall of the stuck pipe are observed at the surface. When the instrument is above the free point, these impulses vary, giving an indication of movement as the pipe is twisted or pulled at the surface. After the instrument passes below the free point, indications of movement cease.

Jet and Chemical Cutters

Stuck pipe may be cut with a *jet cutter,* a wireline tool sometimes used to cut casing in order to salvage it. The jet cutter (fig. 30) contains a shaped charge that explodes outward when detonated from the surface and cuts a thin transverse line through the pipe. The *chemical cutter* (fig. 31) is a more recent innovation than the jet cutter. It shoots out a blast of powerful acid to cut the pipe smoothly, without flare or metal distortion. Neither the jet nor the chemical cutter cause damage to an outer string of pipe.

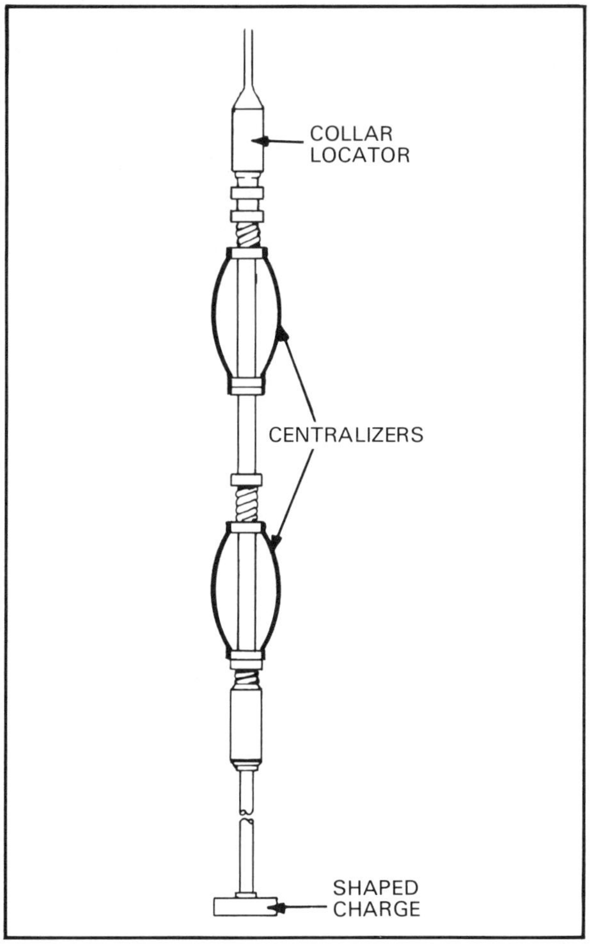

Figure 30. Jet cutter

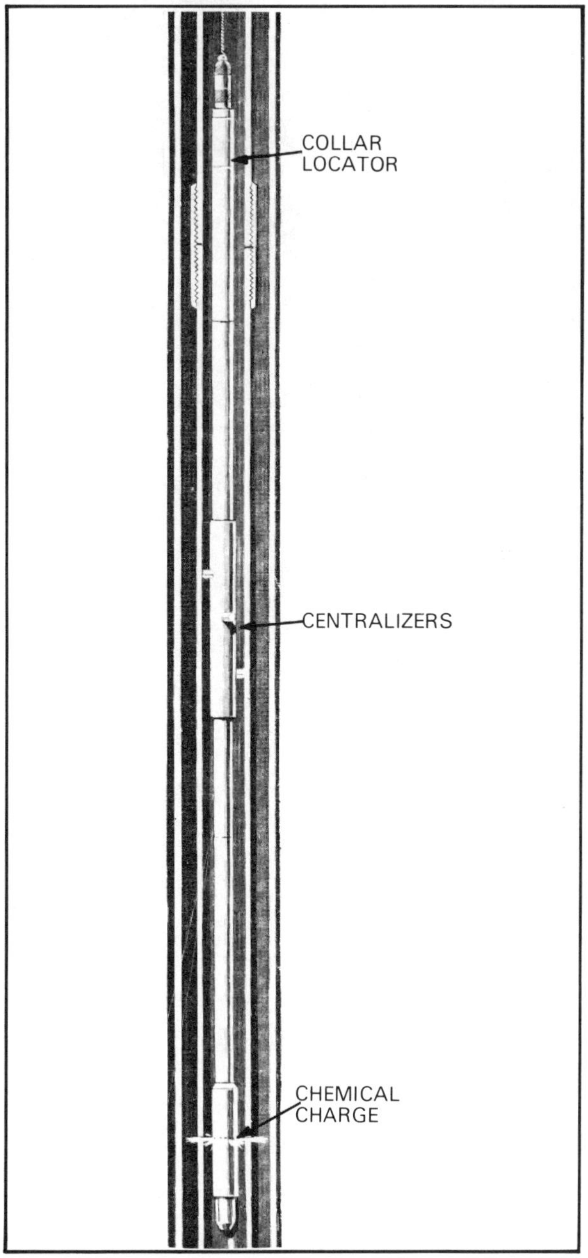

Figure 31. Chemical tubing cutter

String-Shot Backoff

The *string shot* (fig. 32) is a length of Primacord explosive that is electrically detonated downhole. The shot is fired opposite a coupling or tool joint above the point where the pipe is stuck. Left-hand torque is placed on the pipe at the surface and held while the shot is discharged. The explosive force causes the joint to unscrew, and pipe above the disconnected joint can be lifted out of the hole. String shots are often run simultaneously with free-point tools. Assemblies with diameters as small as ¾ inch can be put together.

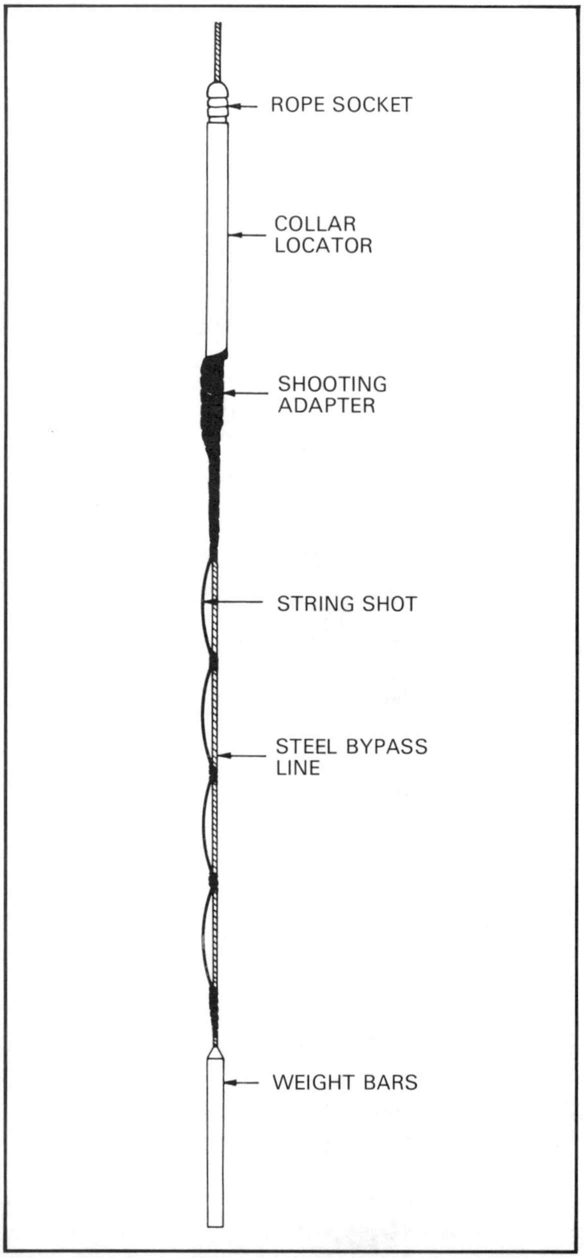

Figure 32. String-shot assembly

REPAIRING AND CONDITIONING TUBING

Scale, rust, bends, and accidental damage can cause obstructions, or burrs, in the tubing string. To reestablish the normal tubing diameter, the sides of the tubing must be forced outward at the point of obstruction so that they are vertical again. Three common mechanical wireline operations, swaging, running a tubing gauge, and running a tubing broach, are used to repair obstructed tubing.

Swaging is an operation in which heavy downward jarring motion is used to force a swaging tool (fig. 33) through the obstructed length of tubing. The diameter of the swaging tool is larger than that of the obstructed tubing. Thus, the sides of the tubing are forced outward without any cutting action.

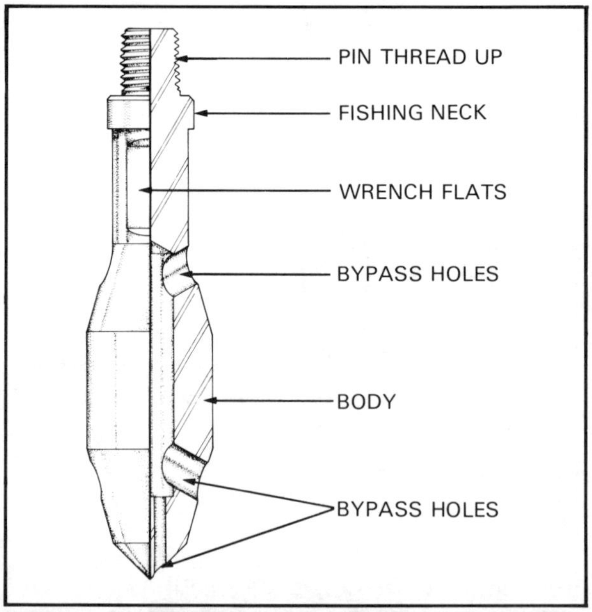

Figure 33. Swaging tool (Courtesy of Camco)

The *tubing gauge* (fig. 34) is sometimes called a *cutter knife* or a *gauge cutter* because of its many sharp bottom edges. The tool is designed to cut away scale, calcium, small deposits of paraffin, and other debris in an obstructed length of tubing. The tool is usually used to measure, or gauge, the internal diameter (ID) of the tubing. The outside diameter (OD) of the gauge cutter is slightly larger than the largest diameter of the wireline assembly that is to be run in the tubing.

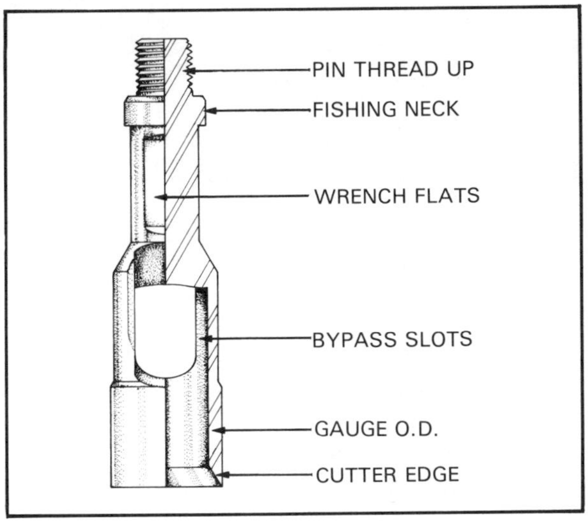

Figure 34. Tubing gauge (Courtesy of Camco)

The *tubing broach* (fig. 35) is functionally similar to the swaging tool in that both are used to force a piece of tubing back to its original diameter. A broach, however, has the capability of cutting metal, if necessary. Operators may use the tubing broach instead of the swaging tool when gradual tubing enlargement is needed. For example, a string of tubing may be lightweight and relatively delicate. Three successively larger spools on the tool are run so that the smallest spool goes through the tubing first and the largest spool, whose diameter corresponds to the tubing's original diameter, goes through the tubing last.

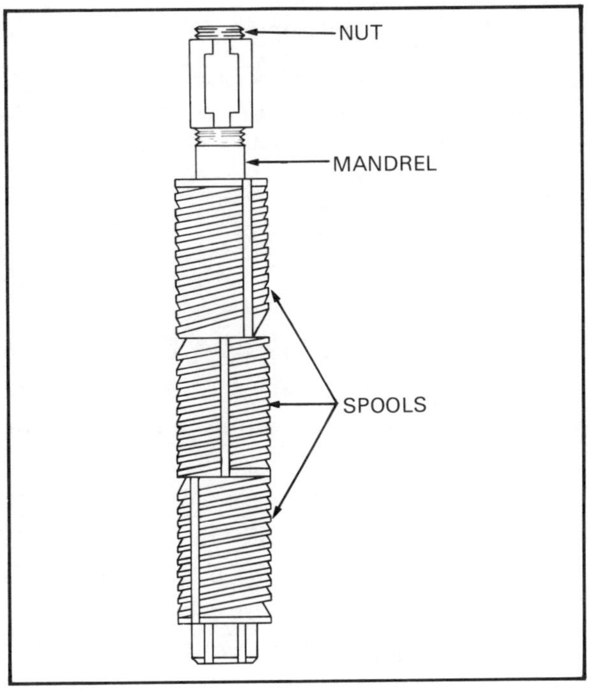

Figure 35. Tubing broach (Courtesy of Otis Engineering)

When a length of tubing develops a hole due to corrosion, drastic bending, or accidental contact with a sharp object, it usually cannot be repaired. Wireline operators may try to solve this problem by avoiding the more obvious and more expensive solution of pulling the tubing and replacing it with a new length. Instead, mechanical wireline is used to lower a tubing stop to just below the tubing leak. Then the stop is set. The stop serves as an anchor for a packoff tool. The *packoff tool* is a length of pipe with sealing elements on its upper and lower ends. Its diameter is smaller than that of the tubing. When heavy downward jarring action is applied from the surface, the sealing elements expand and fit tightly against the inside diameter of the tubing, deflecting rising pay fluids from the tubing and into the packoff tool. The tubing leak is thus bypassed and rendered harmless. Since no two tubing leaks are exactly the same, the length of the packoff tool is custom-fitted to the job. A packoff tool longer than 10 feet is generally run in sections because it is difficult to raise enough lubricator to cover longer tools.

CONTROLLING SAND AND PARAFFIN

Production operators have been plagued for more than a century by the problem of sand produced with oil and water. Despite recent innovations (screen liners, gravel packs, and chemical consolidation treatments) that have helped to lessen the severity of the problem, sand influx remains a problem that requires remedial operations. Sand can build up in the well's perforations and stop the flow of hydrocarbons, particularly in a formation that is relatively loose and unconsolidated. The abrasive sand erodes important subsurface valves and other production equipment, and the gritty substance occasionally forms bridges within the tubing or casing. These bridges may partially block the flow of hydrocarbons and create higher pressure below the bridges than above them, or they may completely block the flow.

Paraffin buildup is a problem similar to sand influx. In some areas, the produced crude oil contains heavy waxes, asphalts, and resins that tend to solidify in the wellbore, forming a sticky substance called *paraffin.* Untreated, a length of tubing affected by paraffin deposits may eventually become completely clogged.

Wireline methods of controlling these troublesome obstructions are used, when needed, to supplement the screen liner, gravel pack, and chemical-consolidation treatments normally associated with major sand control and to provide the primary means of removing paraffin deposits.

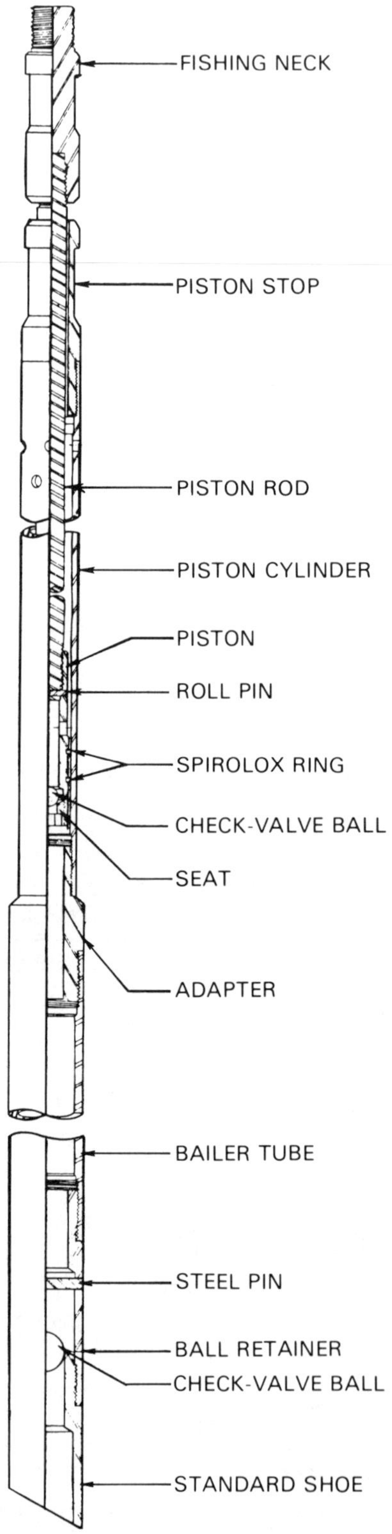

Figure 36. Sand bailer (Courtesy of Camco)

Sand Bailing

Sand may be removed from the wellbore by running a tool called a *sand bailer* (fig. 36) on mechanical wireline. The sand bailer consists of a piston rod, a ball-check valve, a hollow cylinder, and a bailer tube. The cylinder is filled with sand by creating a vacuum within the bailer tube and sucking up sand. Then the bottom of the bailer is closed to trap the sand. If an operator suspects a pressure buildup below the targeted sand bridge, he should equalize the pressure in the wellbore before running the tool. Otherwise, pressure from below the sand bridge can blow the tool string up the hole and result in the wireline tools becoming stuck in the well.

Another type of sand bailer, called a *hydrostatic bailer* (fig. 37), has no piston and requires only a trip to the bottom to pick up a load of sand. The bailer consists of a sealed cylinder with a disk and a piercing tool at the bottom. A sharp blow on the bottom ruptures the disk, allowing the hydrostatic pressure of the well fluid to force sand into the bailer. The hydrostatic bailer is ineffective in wells having very low bottomhole pressure.

Sand Control Guns

An electric-line method of sand control involves shooting a large number of perforations through the casing at the producing formation. The theory behind this practice is that the sand grains form a bridge across the small opening, thereby preventing further movement of sand into the wellbore.

Two types of guns are used for sand control perforating. In both, the charges are exploded to send many small jets into the casing.

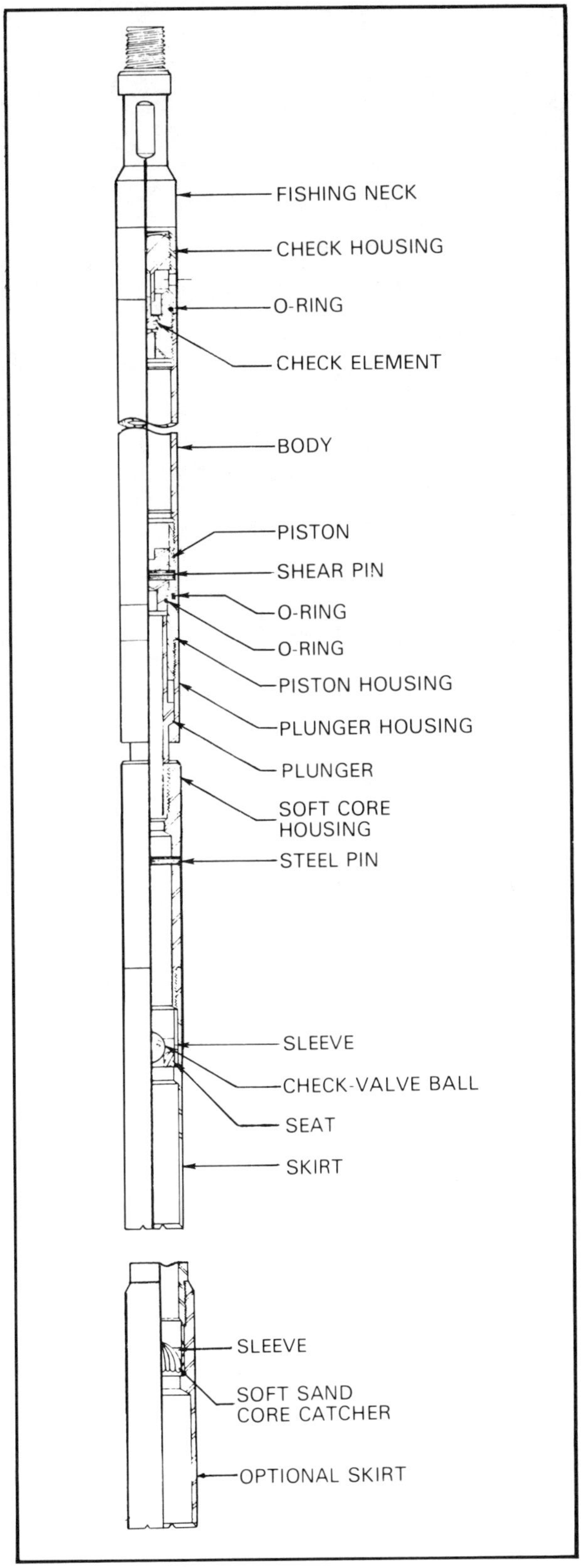

Figure 37. Hydrostatic sand bailer (Courtesy of Camco)

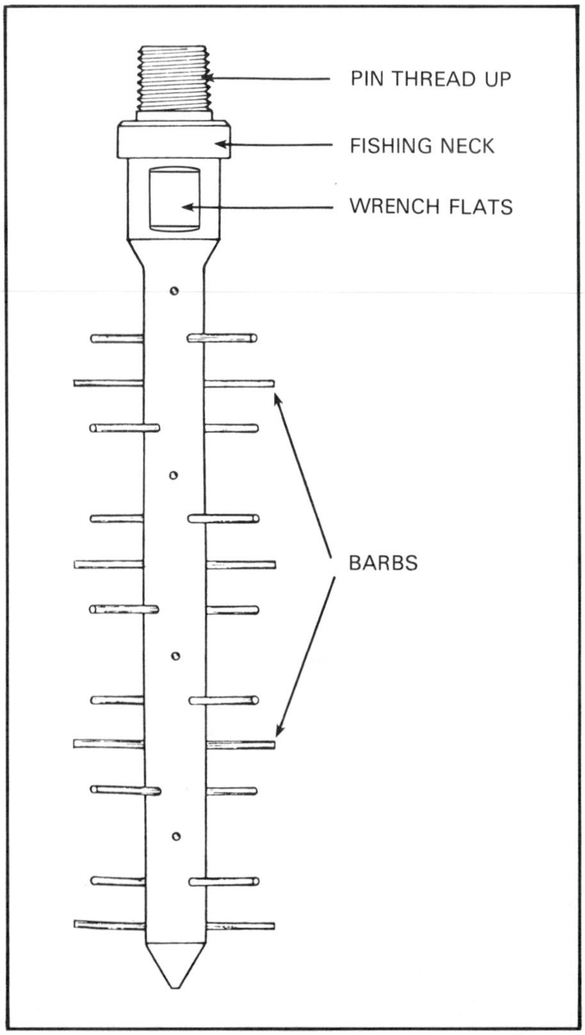

Figure 38. Paraffin scratcher (Courtesy of Camco)

Paraffin Scratching

Paraffin deposits along the inner wall of a tubing string can be scraped off using the tubing gauge. Another way to remove paraffin involves running a paraffin scratcher downhole on slick line. The *paraffin scratcher* (fig. 38) is a long, cylindrical device with barbs that radiate outward in all directions. The tool is normally run while allowing a controlled amount of oil to flow out of the well so that, as the barbs cut loose pieces of paraffin, the paraffin bits are washed up out of the tubing.

FISHING

Wireline fishing, normally performed with mechanical wireline and most often performed during the production phase of the well, is aimed primarily at retrieving small foreign objects such as dropped wrenches and other simple hand tools accidentally dropped downhole by field personnel. In addition, wireline fishing can recover lengths of dropped wireline, as well as tools that have had to be abandoned downhole due to sticking or because of broken line. Most fishing operations are performed on tubing strings, drill strings, and lengths of *spaghetti tubing* (a thin, flexible string of pipe used in a number of wellbore operations). Some of the most common wireline fishing devices include the impression block, the wireline spear, the overshot, the magnetic fishing tool, and the alligator grab.

The *impression block* (fig. 39) is a cylinder filled with soft lead. When the tool is lowered onto a piece of junk, an impression of its size and shape is made in the lead. After reading the fish's signature, the operator can then decide on the specific tool needed to retrieve it.

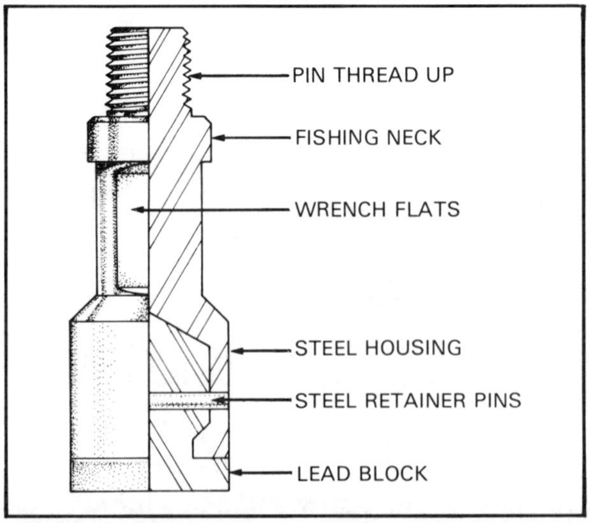

Figure 39. Impression block (Courtesy of Camco)

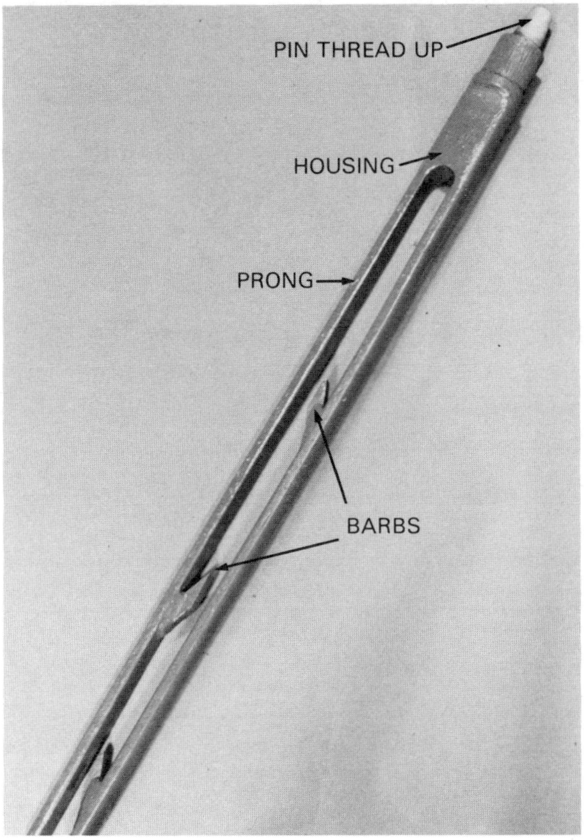

Figure 40. Wireline grab

The *wireline grab* is a particularly ingenious device designed to catch a length of broken wireline in the hole and bring it to the surface. It consists of a long, thin U-shaped housing with two or three barbed prongs welded to the inner sides (fig. 40). Once a coil of wireline is trapped in this device, it can not easily fall back downhole.

The *alligator grab* is used to pick up relatively small objects like wrenches that have fallen or have been accidentally dropped into the wellbore. It is designed with jaws like an alligator's. The jaws are pinned open before the grab is run into the well. Then, when contact with the fish is made, the jaws snap shut and are held shut by the tension of a coiled spring within the tool.

The *magnetic fishing tool* is, as its name implies, a magnetic tool used to fish from the well any small metal object, a tiny steel cutting for example, that could be attracted to a magnet.

Self-Test

4: Troubleshooting Operations

(Multiply each correct answer by five to arrive at your percentage of competency.)

Fill in the Blank

1. List the four most commonly requested wireline troubleshooting operations.

   (1) ______________________________

   ____

   (2) ______________________________

   ____

   (3) ______________________________

   ____

   (4) ______________________________

   ____

2. List three methods of freeing stuck pipe.

   (1) ______________________________

   (2) ______________________________

   (3) ______________________________

3. A free-point indicator uses ____________ ____________ to indicate movement of pipe above the free point.

4. The ____________ ____________ contains a shaped charge that explodes outward when detonated from the surface and that cuts a thin transverse line through pipe.

5. The ____________ ____________ shoots out a blast of powerful acid to cut pipe smoothly.

6. A ____________ tool is forced through tubing with a heavy jarring motion to force the sides of obstructed tubing outward without any cutting motion.

7. The ________________ ________________ is used to measure the internal diameter (ID) of tubing and to cut away scale, calcium, paraffin, and other debris in an obstructed length of tubing.

8. The ______________ __________________ consists of a piston rod, a ball-check valve, a hollow cylinder, and a bailer tube.

9. The ____________________________ ______________ __________________ consists of a sealed cylinder with a disk and a piercing tool at the bottom.

10. An electric-line method of sand control involves shooting a large number of ____________________________ through the casing at the producing formation.

11. List the five most commonly used wireline fishing devices.

(1) ____

(2) ____

(3) ____

(4) ____

(5) ____

5

Completion and Production Maintenance Operations

OBJECTIVES

Upon completion of this section, the student will be able to:

1. Describe the three types of completions.
2. Describe the advantages of the perforated completion over the other two types of completions.
3. Describe the three methods of perforating and their uses.
4. Describe how a well is swabbed.
5. List the subsurface wireline equipment included as part of the tubing string.
6. List the flow control equipment used in a wireline-completed well.

WELL COMPLETION

Well completion refers to the series of operations that transform a drilled and tested borehole into a producing oilwell or gas well. A wide array of completion methods and techniques has been developed in order to–

1. allow the pay fluids to channel horizontally into the hole after drilling has tapped the producing formation and
2. allow operators to predict and control the upward flow of oil and gas into the borehole.

Early Completion Methods

Early operators completed a well by setting a string of casing to the top of the producing formation and allowing the well to produce into the open or uncased hole. This open-hole, or barefoot, completion did not provide for control over sand influx nor for placement of stimulation treatments, and it is rarely used today. A second completion method, the liner completion, was also developed early in the industry's history. A screened liner, a pipe with holes or slots that is suspended from the casing, was run to the bottom of the wellbore. Again, however, operators who used a screened liner were left with little control over which intervals of the reservoir contributed to production; that is, they had no control over water production from below the oil-bearing part of the zone.

By the mid-1930s, the industry, in response to the twin problems of horizontal access and production efficiency, developed the perforated completion, which is used by the majority of today's operators.

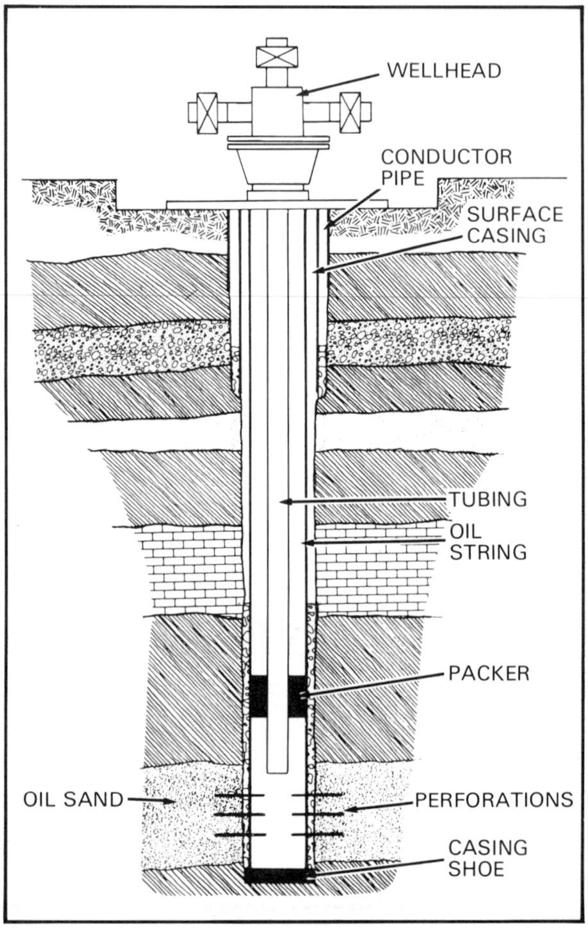

Figure 41. Single-zone perforated completion

Perforated Completions

In a perforated completion, a string of casing is run and cemented across the productive zone. The casing is then perforated at precisely determined depths opposite the zone, creating channels through which fluids can enter the wellbore and can be brought to the surface (fig. 41). Perforated completions are widely used because they can be adapted to all kinds of formations, including those with sand control problems, and because pay zones can be selectively stimulated and perforated. The perforation method also allows more than one zone to be produced (fig. 42). However, most oilwells and gas wells in the United States

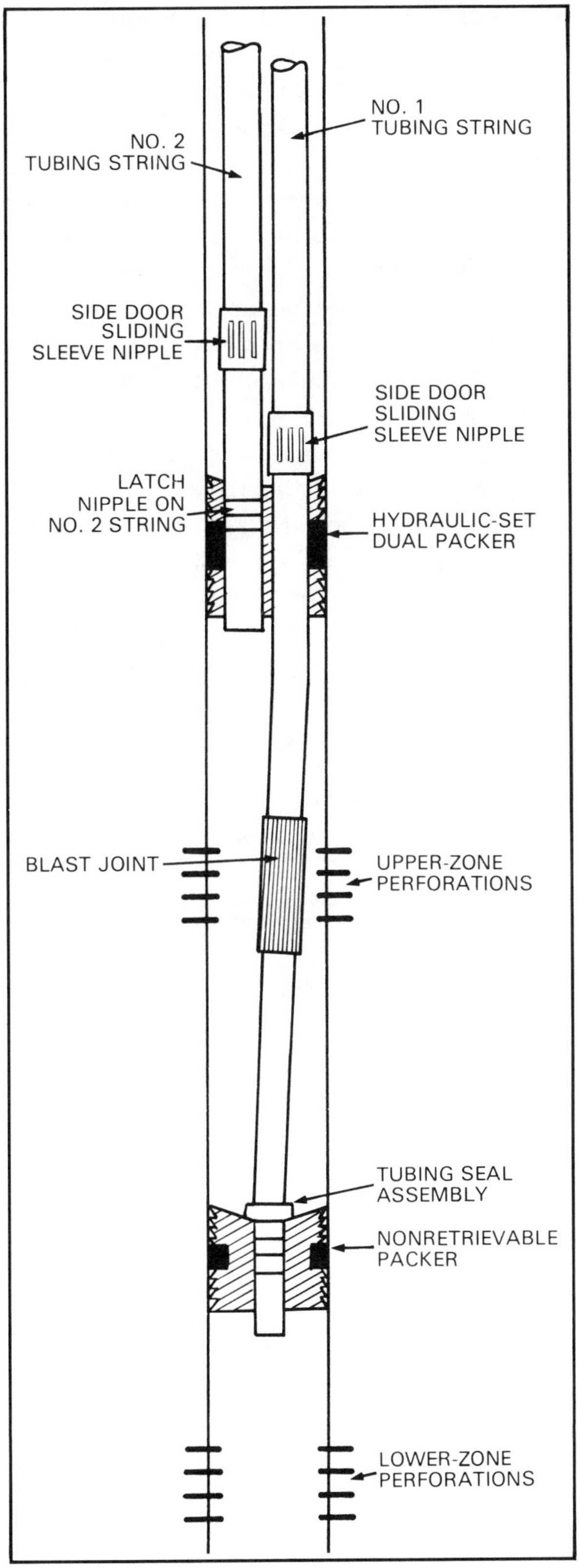

Figure 42. Multiple-zone perforated completion

(about 90 percent) are completed as single-zone producers. The majority of these wells are single-zone perforated casing completions. The primary advantage of a perforated completion is that it isolates the zone to be produced and prevents the flow of other fluids such as water in a well.

Tubing, which is a string of small-diameter pipe, increases the efficiency of production in several basic ways. The safety of a completion is increased because fluid may be circulated or swabbed if necessary to bring the well in. Tubing also provides a way to circulate fluid to kill the well when the need arises. Also, tubing can be pulled if it becomes plugged or damaged. Finally, when used in conjunction with a *packer*—a tight seal between the tubing and casing that is set above the producing zone—tubing keeps well fluids away from casing and thus protects the larger pipe from corrosion.

PERFORATING

The first step in the perforating operation is to lower a perforating gun on wireline to a specific location in the wellbore and to fire it either electrically or by mechanical jarring action to create holes in the casing, in the tubing, or in both simultaneously. The three common methods of perforating are jet perforating, bullet perforating, and mechanical perforating. Both jet perforating, which is also called shaped-charge perforating, and bullet perforating open channels in the productive formation for a perforated completion. Mechanical perforating, on the other hand, is used to punch holes in the tubing to circulate or to kill a producing well.

Jet Perforating

Jet perforating shoots one or more shaped charges through casing and cement and into the producing formation. Developed during World War II to penetrate tanks and later adapted for use in the oil field, the shaped charge is hollowed out at its forward end and lined with metal in the hollow part to increase its penetrating power. These small but powerful charges are fired for 90 to 95 percent or more of today's perforated completions.

A perforating crew usually consists of an engineer and two operators employed by a cased-hole wireline company or by the cased-hole division of a large firm like Schlumberger or Dresser Atlas. When a crew is called for a perforating operation at a wellsite, their first job is loading the charges into a hollow, rugged carrier gun. The charges are positioned in a vertical spiral around the gun (fig. 43) so that holes can be punched in all directions along the desired formation. After the gun is loaded, the engineer checks out the complex wiring connecting the shaped charges to the electric wireline. Both of these tasks are performed in the shop. Once the crew is in the field, such preliminaries only cost the operator rig time. Most perforating tools are far too complicated to fix at the wellsite.

The operator must inform the wireline service company of casing size, whether the hole is empty or filled with fluid, and the expected well pressure when ordering a perforating job. Casing strings have different sizes of internal diameters (ID), and tools must be chosen accordingly. A fluid-filled hole both cushions the sensitive tools and provides positive well pressure to help offset a possible perforation-induced surge in bottomhole pressure. When such a surge is possible, a condition of negative pressure exists. The perforating crew must plan to install pressure containment equipment on

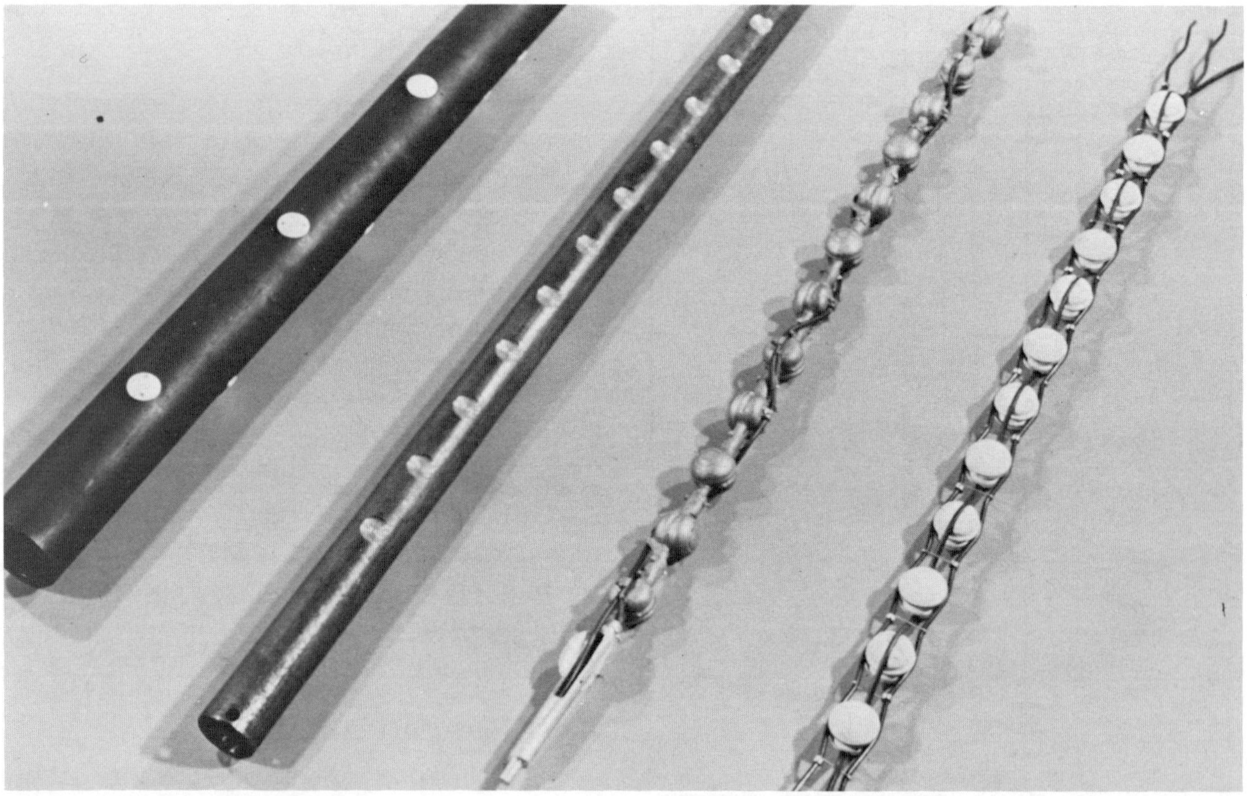

Figure 43. Jet perforating guns

top of the wellhead before running the gun. For this reason, perforating under negative pressure is usually done through tubing after the Christmas tree has been installed. If the hole is empty, perforating almost always results in sudden high-pressure flow toward the surface; an empty hole can also cause the gun to kick vertically.

After the gun has been loaded into the wireline truck and the crew has arrived at the wellsite, the operators assemble the tool string while the engineer calibrates the equipment in the cab of the truck. To ensure that the casing is perforated at the exact desired depth, gamma ray or collar logging tools are made up with the guns. Their readings are correlated with earlier logs provided by the operator. When the tool string has been made up and the equipment in the cab has been checked out, a hoisting rig, either provided by the operator or brought along by the wireline crew, lowers the tool string into the hole. When the desired depth has been reached, the engineer flips a switch in the cab of the wireline truck and detonates the charges.

Bullet Perforating

While jet perforating is the method of choice in the vast majority of drilled formations, bullet perforating is another alternative in formations having low to moderate compressive strength. Widely used before the development of jet perforating, bullet perforating is only rarely used today. Due to the temperature limitations of the explosives involved, bullet perforating is not used in a well with a temperature higher than 250°F.

A bullet perforator is a multibarreled gun. The charges are usually pistol shells that can be selectively fired. The combustion chamber is in the shell, and each bullet has its own barrel. Like the jet perforator, the bullet perforator is made up by the wireline crew in the shop, carried to the wellsite, assembled on a tool string, and lowered to the desired depth with a hoisting rig. Again, a switch in the cab of the wireline truck fires the gun and shoots the bullets into the formation (fig. 44).

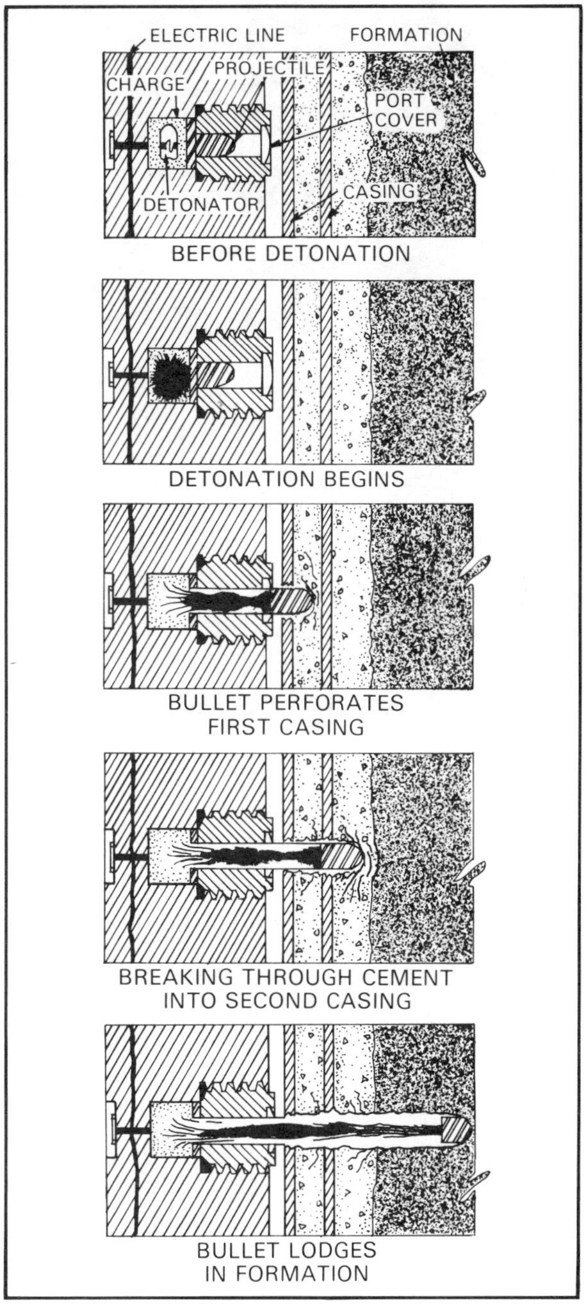

Figure 44. Bullet perforating operation

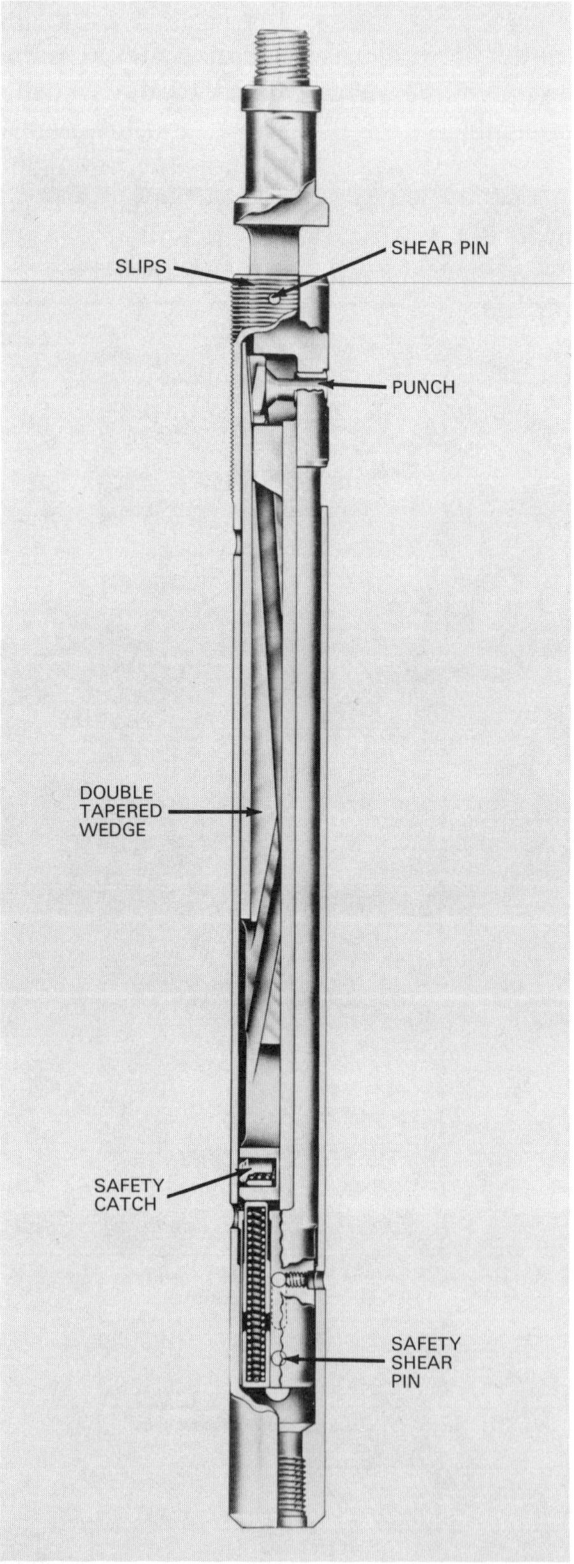

Figure 45. Tubing perforator (Courtesy of Otis Engineering)

Mechanical Perforating

A common mechanical perforating procedure is used to punch holes, one at a time, into the tubing string to circulate or to kill a well. An operator uses this operation when tubing-casing communication is needed in a production string not equipped with permanent circulating devices. The operator may, for example, want to mix a certain chemical into the pay fluid. With tubing perforations, a precisely controlled amount of the chemical can be pumped into the tubing-casing annulus so that it travels through the hole and into the producing fluid.

A tubing perforation operation is performed under pressure using standard wireline procedures and hydraulic jars. The perforating tool at the end of the string consists of a hollow cylinder containing a powder charge (fig. 45). Hydraulic jarring action imposed from the surface fires the powder charge, and expanding gas from the explosion drives a wedge down behind a small button and pushes the button out through the tubing. While the force of the perforation allows tubing penetration, it is not sufficient to affect the casing. The button simply plops harmlessly against the casing wall and drifts down the annulus to the bottom of the hole.

SWABBING

Sometimes the bottomhole formation pressure in a newly completed well is not sufficient to lift and move out the drilling fluid still in the hole. *Swabbing* is a procedure used to raise the drilling or completion fluid to the surface and thus to lower the pressure inside the hole and to allow the higher formation pressure to send the pay fluids through the perforations and up the tubing. Swabbing is sometimes referred to as *unloading the well.*

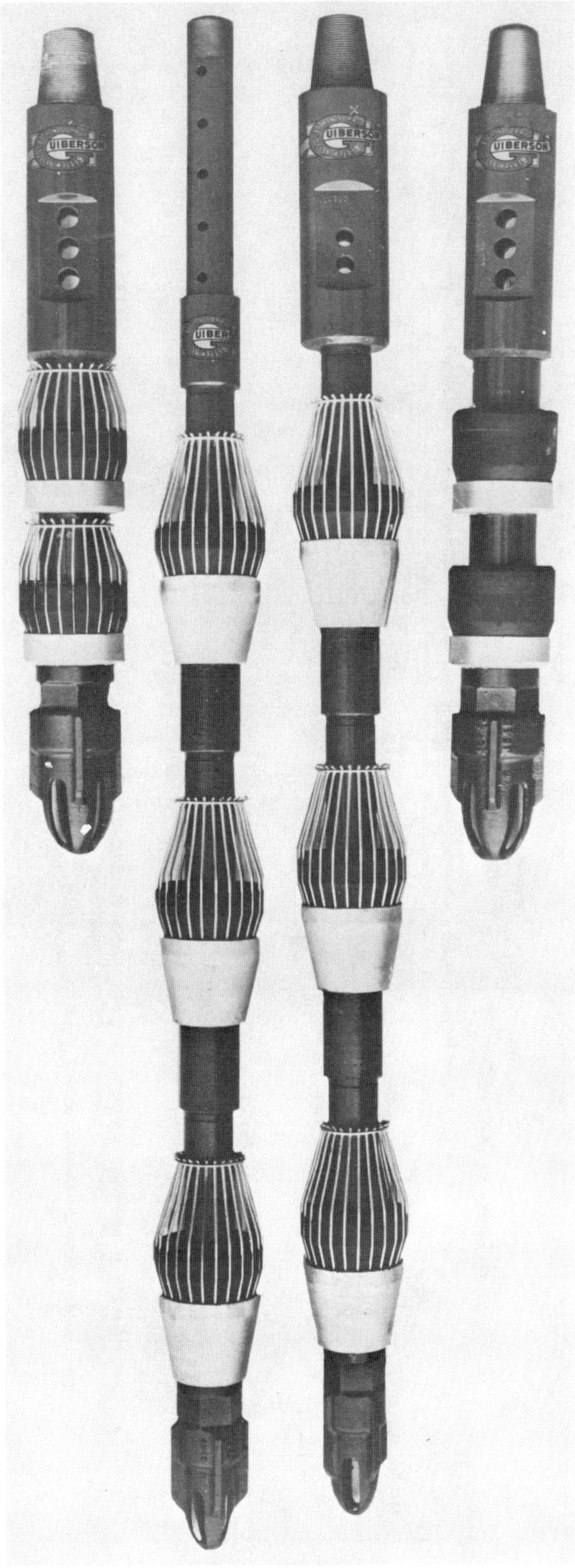

Figure 46. Swab cups

When a well is particularly deep and the drilling fluid is extremely heavy, swabbing is performed using a regular workover rig. The swabbing equipment is lowered and raised with sand line, which is sturdier than wireline and capable of raising longer and heavier columns of fluid to the surface. When a lighter job is called for, however, the well owner calls in a wireline crew. While wireline swabbing is limited in the amount of fluid that can be raised in one trip – up to 300 column feet, depending on fluid density – the wireline alternative is, again, quicker and less expensive than a workover job.

Wireline swabbing usually occurs after the Christmas tree has been installed. Using standard mechanical wireline procedures, one or more concave, rubber-faced discs called *swab cups* (fig. 46), are lowered into the wellbore. A swab cup is almost as large in diameter as the tubing. When the swab cup is raised, a lip rimming the outer edge of the cup seals against the tubing ID and forms a tight bottom seal under the rising column of fluid. The swab cup functions like a piston. If too much fluid has been picked up and its weight threatens to snap the line, the lip collapses, and the fluid is dumped back down the hole.

PRESSURE AND FLOW CONTROL

Once the casing has been set and perforated to allow for horizontal channeling of the oil or gas, a production string that gives maximum control over the pay fluid's pressure and its rate of upward flow is made up and set. When conditions permit, a tubing string is installed so that wireline servicing is possible. Wireline services provide versatility, convenience, and low expense not provided by alternate methods.

Permanent Wireline Tubing String

The most widely used completion is a permanent tubing string installed on a single-zone, perforated well. This completion, which may be serviced with wireline, uses a packer to isolate the tubing from the casing and to keep formation pressure and fluids out of the casing. Downhole wireline equipment is also installed on wells completed without a packer, as well as on dual- and multiple-zone producers. The permanent tubing equipment and temporary flow and pressure control devices used on these less common completions remain, however, essentially the same as for the more common completion using a packer. The usual subsurface components installed in the tubing during wireline completion procedures include selective and no-go landing nipples, flow couplings, and circulating devices.

Landing Nipples. Run as an integral part of the tubing string, a landing nipple provides a seat at a known depth into which a wireline crew can set various types of retrievable flow control equipment. The landing nipples are machined internally to receive the mandrels, or removable locking devices, used to position, lock, and seal the subsurface production controls in the tubing. An operator planning for a wireline completion should consider the uses and downhole placement of both selective and nonselective landing nipples.

One commonly used nipple (fig. 47) has a selective internal profile machined to accept only a locking mandrel that has a fixed external profile keyed to a specific nipple. As many as five selective nipples may be included in the tubing string. They must be run in sequence with the nipple with the smallest profile nearest the well's bottom. The sequence of the nipples and their depths must be recorded in the well files.

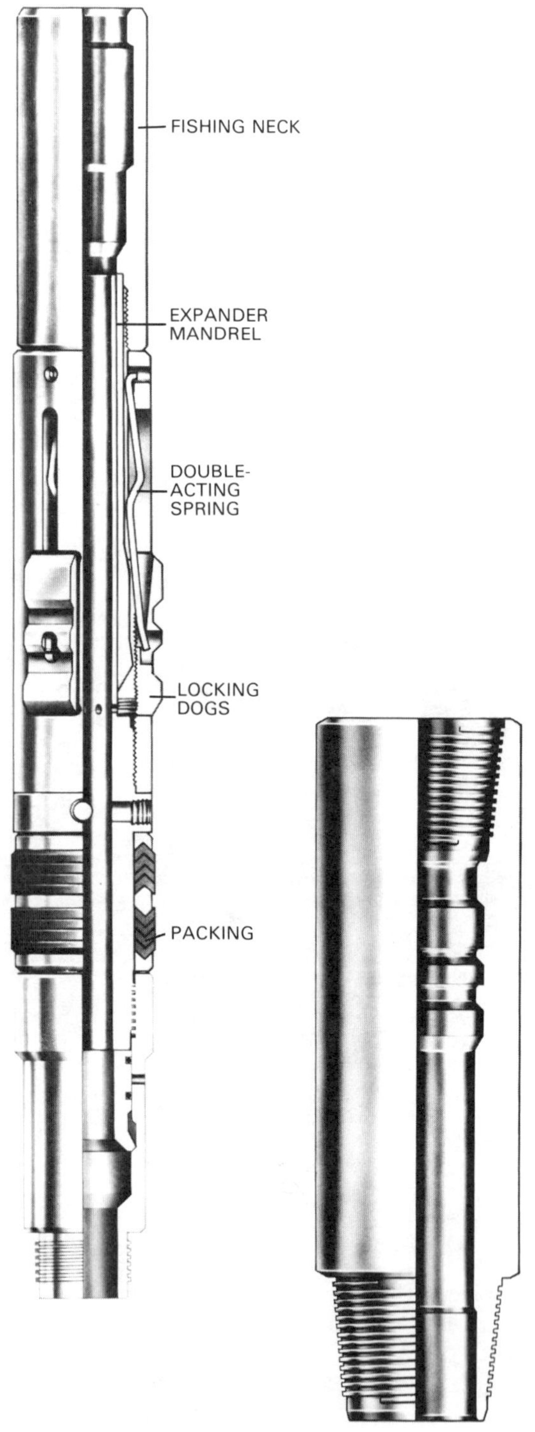

Figure 47. Selective landing nipple and locking mandrel (Courtesy of Otis Engineering)

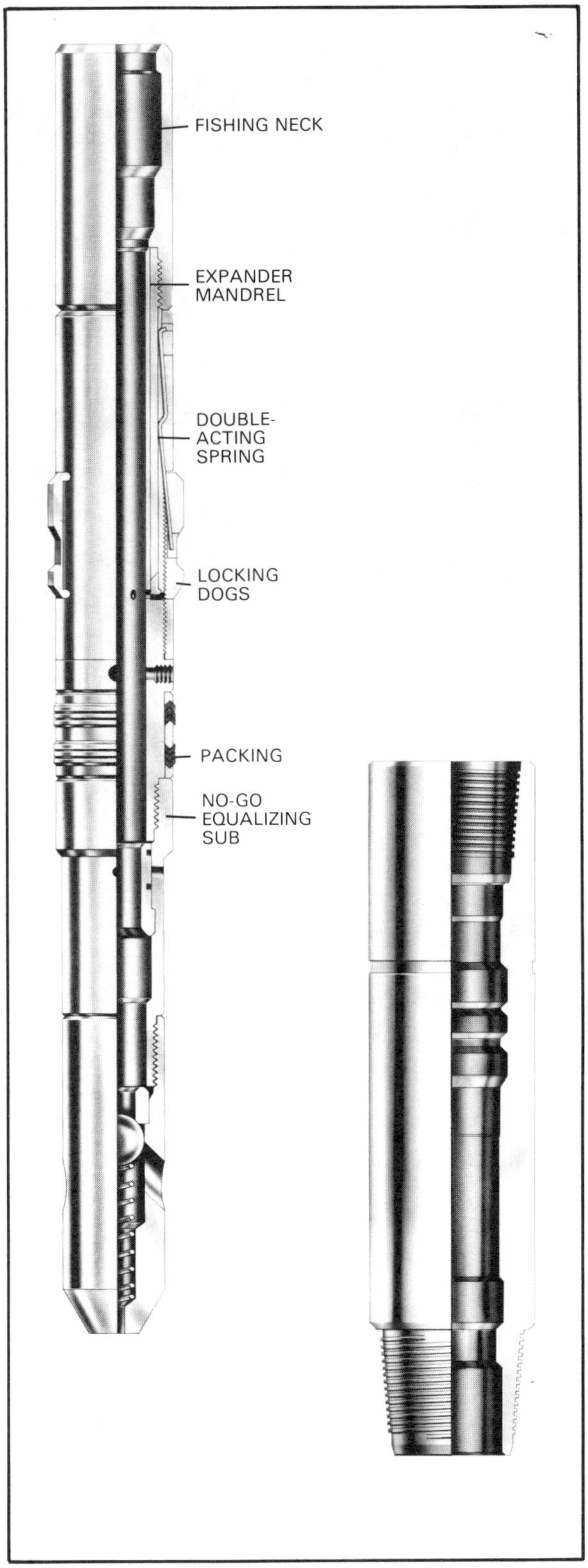

Figure 48. No-go landing nipple and locking mandrel (Courtesy of Otis Engineering)

When a given flow control device is being set at the depth of a particular nipple, the piece of equipment can be lowered into the hole with a locking mandrel that bypasses the larger nipples and lands in the correct one.

The universal nipple can be placed in the tubing string at any position and at as many locations as may be desired. Universal nipples are all identical. The running tool selects the nipple in which the locking device is set.

A nonselective, or no-go nipple (fig. 48), is a restricted ID landing nipple normally placed at or near the bottom of the tubing string. Only one no-go nipple of a given size can be placed in a string. Locking tools corresponding to the no-go nipple bypass selective nipples above the no-go nipple before the locking tool reaches the restricted ID at the lower end of the no-go nipple and stops.

Landing nipples are used to perform many vital procedures in a single-zone well. Some of these are setting and retrieving the tubing safety valve, plugging the tubing from above or below, and hanging a bottom-hole tool or sand screen.

Flow Couplings. Great turbulence can result from the placement of flow control devices in landing nipples. Flow couplings are placed immediately above, and sometimes below, the nipples to protect the tubing from possible erosion. Flow couplings are normally about two feet long and are made of high-grade alloy steel; their ID equals that of the tubing.

Circulating Devices. The circulating device can be manipulated with wireline to establish communication of fluids and pressure between the tubing and the tubing-casing annulus. This communication is often needed to kill a well or to treat the pay fluid with chemicals.

The sliding sleeve valve (fig. 49), is one type of circulating equipment. This device is opened by downward jarring and closed by upward jarring after it is engaged by a positioning tool.

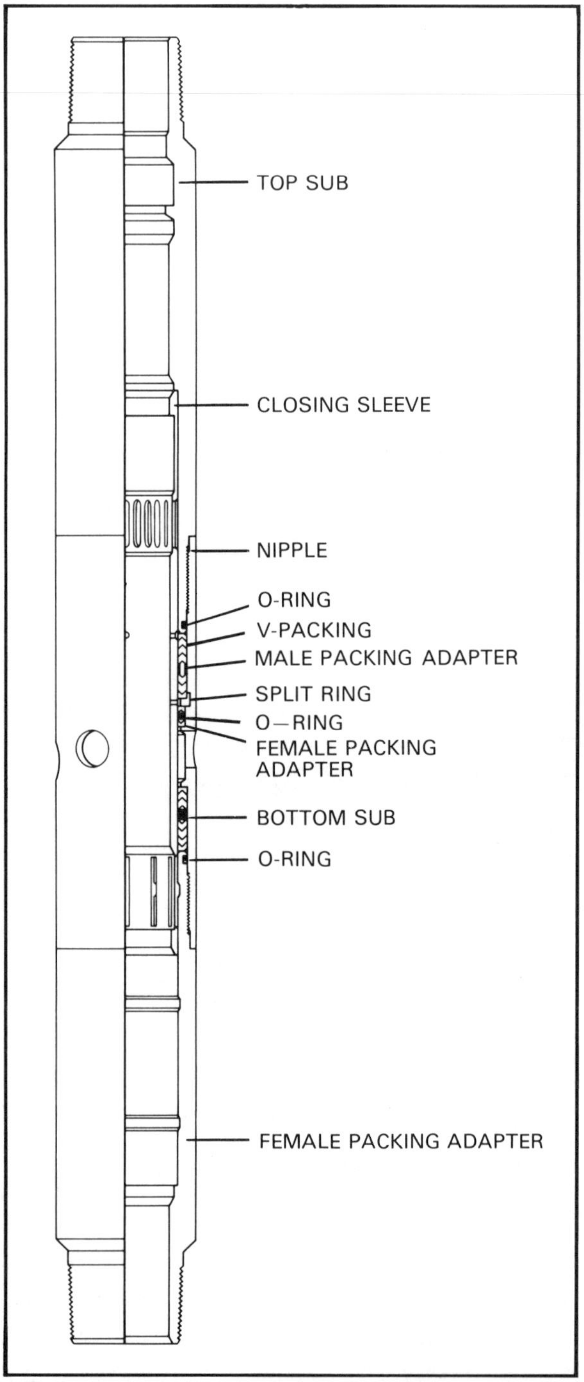

Figure 49. Sliding sleeve valve, or sliding side door (Courtesy of Otis Engineering)

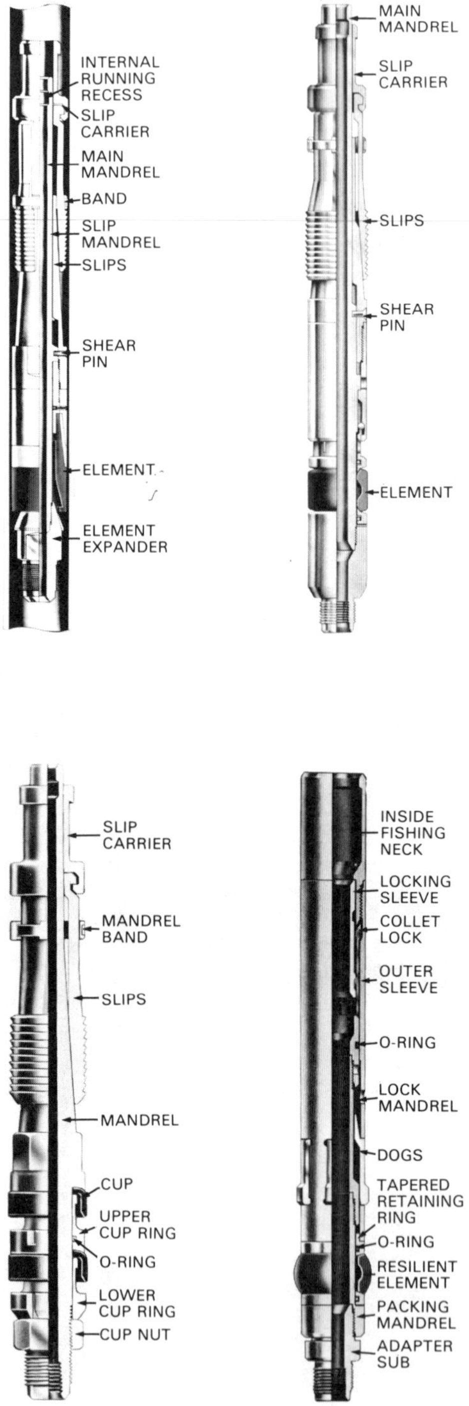

Figure 50. Lock mandrels designed to lock and pack off wireline retrievable flow controls in wells not equipped with landing nipples.

Mandrels. *Locking mandrels,* also called *locks* and *mandrels* (fig. 50), are used to hold flow controls in place against the upward thrust of the flow of the well. Locking mandrels fall into two categories: tubing locks and nipple locks. Tubing locks are used to secure a subsurface flow control in a well completed without landing nipples. These locks are generally not capable of working against differential pressure over 1,750 psi. Nipple locks are located in a nipple in the tubing string. The nipple is designed so that the lock may be set and retrieved using wireline.

Side Pocket Mandrel. The side pocket mandrel (fig. 51) is made up in the tubing string as the tubing is being run into the well. Later, a gas-lift valve may be run into the well on a wireline and placed in the side pocket of the mandrel. Instead of having to pull the tubing to repair or replace the valve, as is the case with the conventional mandrel, the side pocket mandrel allows valves to be installed or removed by wireline. Side pocket mandrels can also be used as receivers for dummy valves, circulating valves, and chemical injection valves.

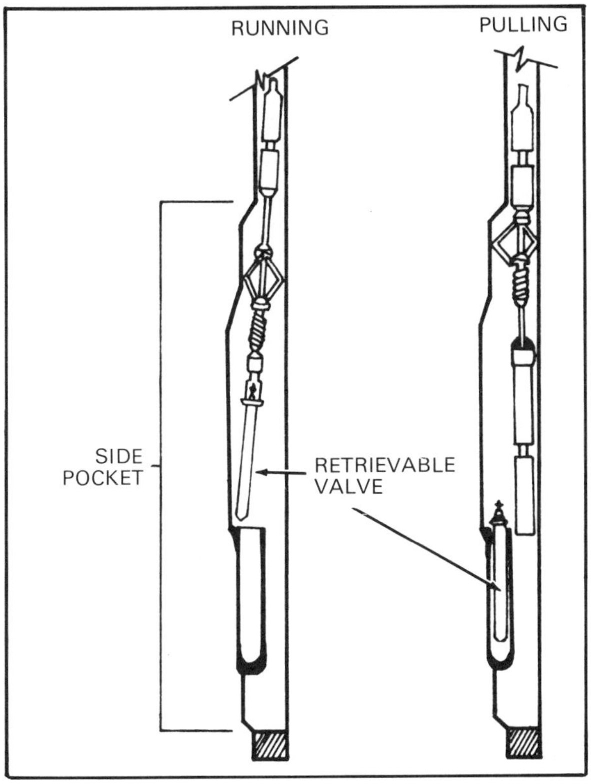

Figure 51. Placement of a retrievable valve in a side pocket mandrel

A *kickover tool* (fig. 52) is used to position wireline devices in a side pocket mandrel.

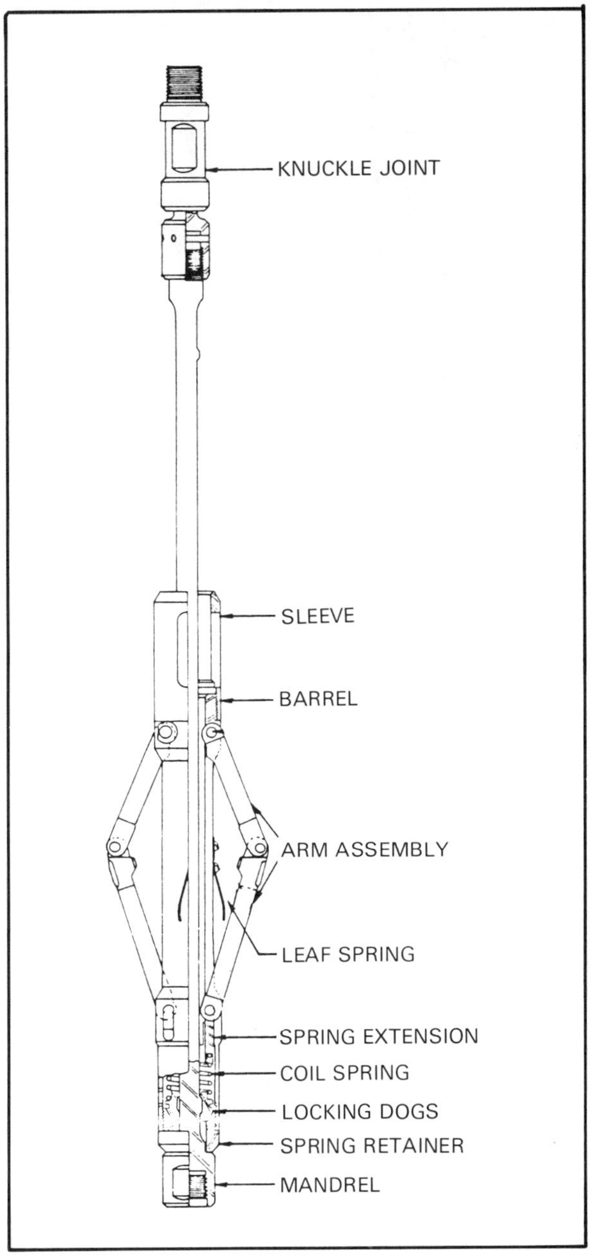

Figure 52. Kickover tool (Courtesy of Camco)

When the tool string assembly reaches the side pocket mandrel, the kickover tool pivots (or kicks over) and places the flow control device in position above the side pocket mandrel.

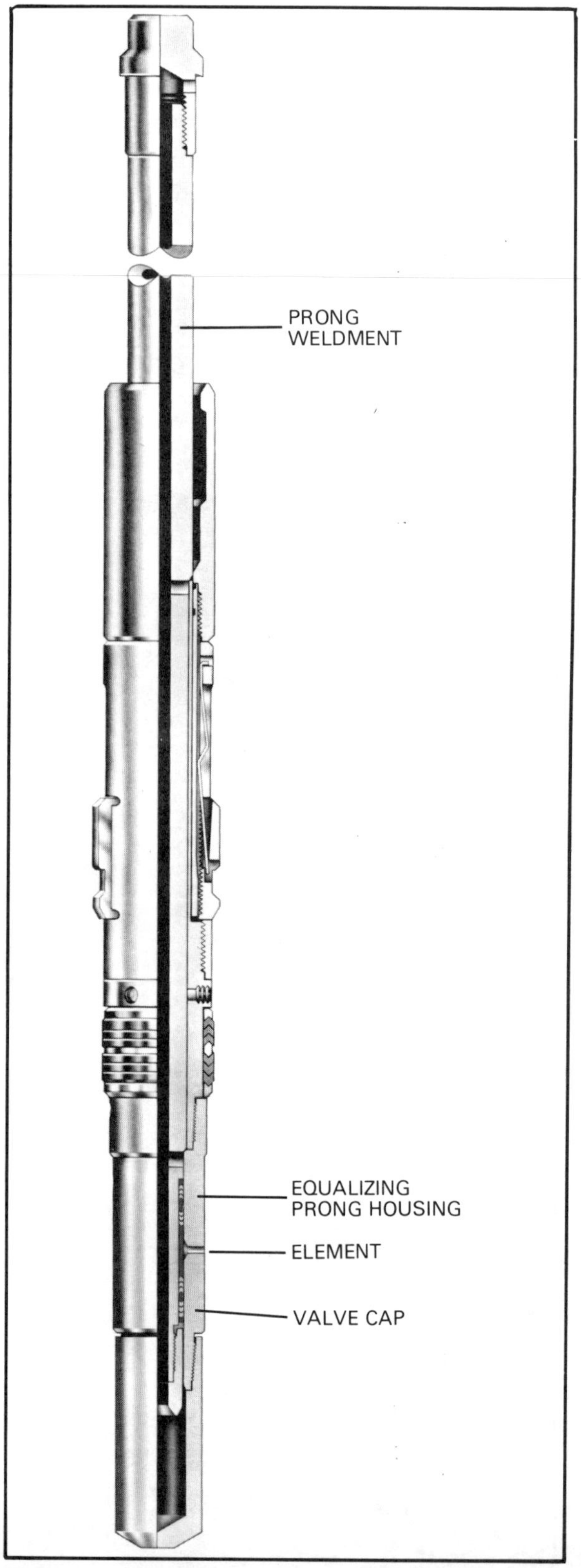

Figure 53. Tubing plug (Courtesy of Otis Engineering)

Setting Subsurface Equipment

After the permanent wireline tubing string has been installed, the operator can call in a wireline crew to perform a wide array of downhole operations designed to control the well's pressure and flow. These operations are performed under pressure, using special tools to set and retrieve the production control equipment and to perform thru-tubing plug-backs and perforating.

A full record of the well's initial completion and subsequent history is requested as the wireline crew prepares to set a downhole production control device such as a tubing plug (fig. 53). After determining the depth at which the device is to be set, the crew then chooses a locking mandrel machined to complement the internal profile of the nipple nearest that depth, attaches it to the control device, and runs the entire tool string into the well. The locking mandrel is activated by surface manipulation of the wireline to go into the nipple and to effect an extremely tight lock of the control device within the tubing. The running tool is then released and the wireline tools are retrieved.

A running tool (fig. 54) is used to install flow control devices in the tubing string. Some running tools use a downward jarring motion, some use an upward jarring motion, and some use both an upward and downward jarring motion to set the lock mandrel. Running tools may be used to set flow control devices in landing nipples or when no landing nipples have been placed in the tubing string.

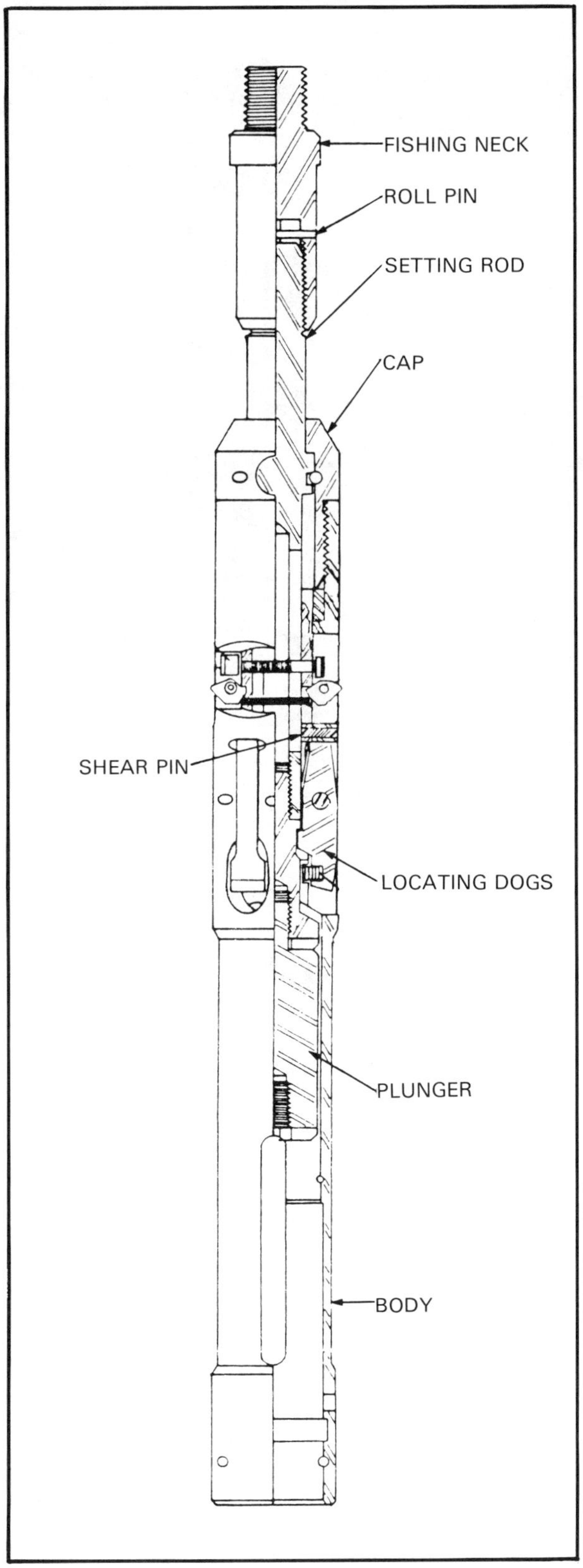

Figure 54. Running tool (Courtesy of Camco)

The wireline crew runs a *pulling tool* (fig. 55) into the well to remove retrievable subsurface devices. Standard pulling tools are machined or built to latch onto an internal or external fishing neck on top of the the subsurface equipment to be retrieved. They may likewise be released from the equipment should the latter become stuck in the tubing.

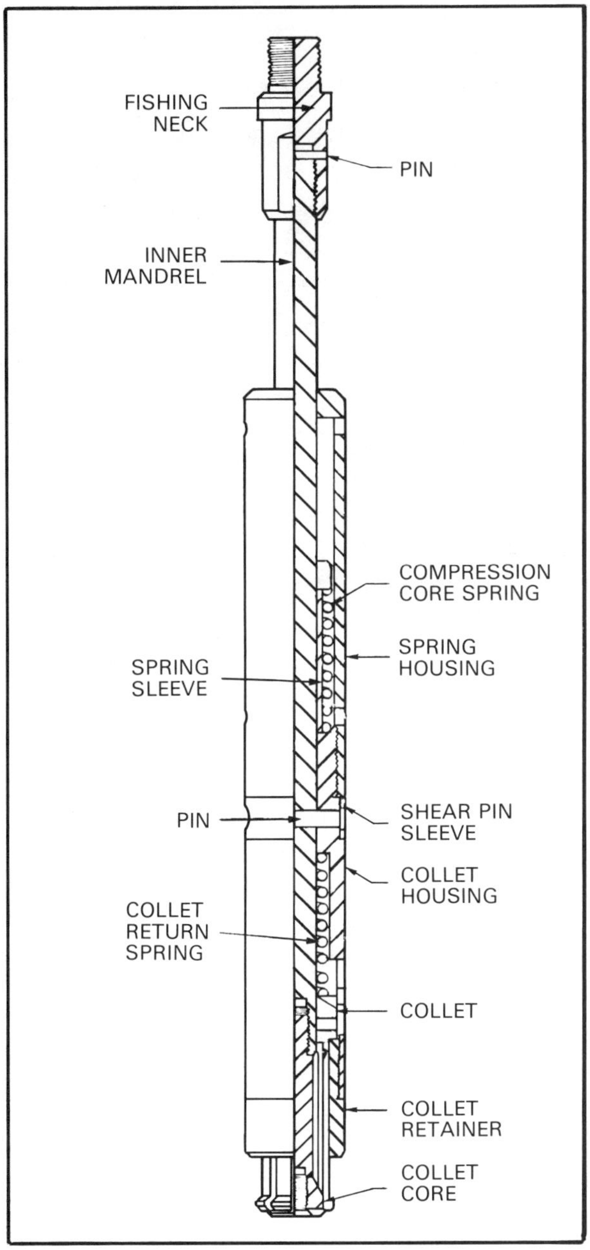

Figure 55. Pulling tool (Courtesy of Camco)

A special sliding sleeve positioning tool is used to move the inner sleeve of common circulation devices. When run to close the sleeve, the positioning tool automatically releases when the circulating port is completely shut. Run in an inverted position, the tool releases when the sleeve is open. A sliding sleeve positioning tool also has a shear pin that permits release in case junk makes it impossible to shift the sleeve.

Subsurface Control Devices

Some downhole equipment commonly used for production and flow control in a wireline-completed well are tubing safety valves, flow regulators, bottomhole regulators, and plug chokes.

Wireline retrievable tubing safety valves (fig. 56) are designed to shut in a well automatically when surface valves fail or are damaged. Safety valves are controlled either at the surface or at their location downhole. The surface-controlled variety is operated by hydraulic pressure in small lines that are run outside the tubing from a control device at the surface. The subsurface-controlled valve has either a spring or a bellows. These valves are required by the American government on all offshore wells in U.S. waters.

Flow regulators are designed to reduce flowing wellhead surface pressure and lessen the possibility that surface controls may freeze up when pressure is lowered at the surface. Wireline retrievable flow controls can be, but aren't limited to, plugs (sometimes referred to as tubing plugs or plug chokes), regulators (a regulator simply reduces the pressure in the tubing downstream of its point of installation), downhole chokes (these devices regulate the flow of well fluids), and subsurface safety valves (these devices close off the flow only if abnormal operating conditions are detected). *Bottomhole regulators* (fig. 57) act as a bottomhole choke to reduce flowing pressure in the tubing. Comprised of a removable valve and spring-loaded seat, the

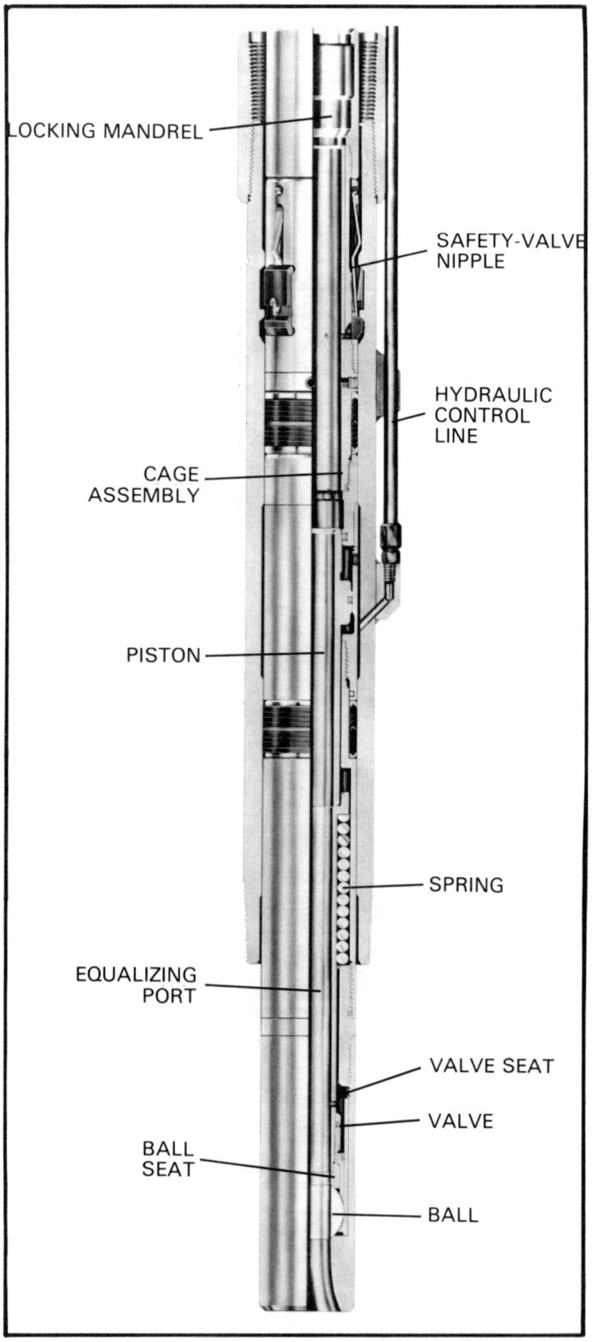

Figure 56. Retrievable tubing safety valve (Courtesy of Otis Engineering)

bottomhole regulator allows pressure reduction to take place at high reservoir temperatures before the fluids are cooled by their trip to the surface. It also helps maintain a lower surface pressure.

Plug chokes are used to plug the tubing at some point along the string (fig. 58). They are set to test for packer leakage, to stop production when wellhead equipment must be removed, to test tubing for corrosion or other damage, and to separate zones for production or stimulation treatments. Some plugs hold pressure from both above and below; others, known as pump-through plugs, permit drilling fluid to be pumped down the tubing to kill the well.

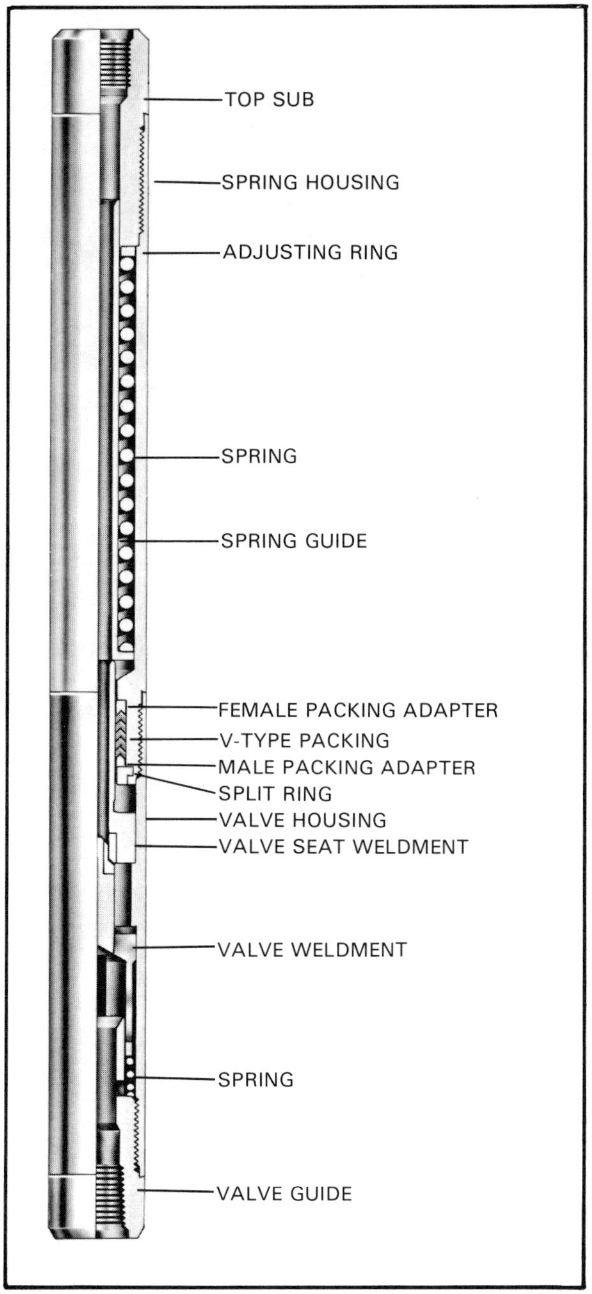

Figure 57. Bottomhole regulator (Courtesy of Otis Engineering)

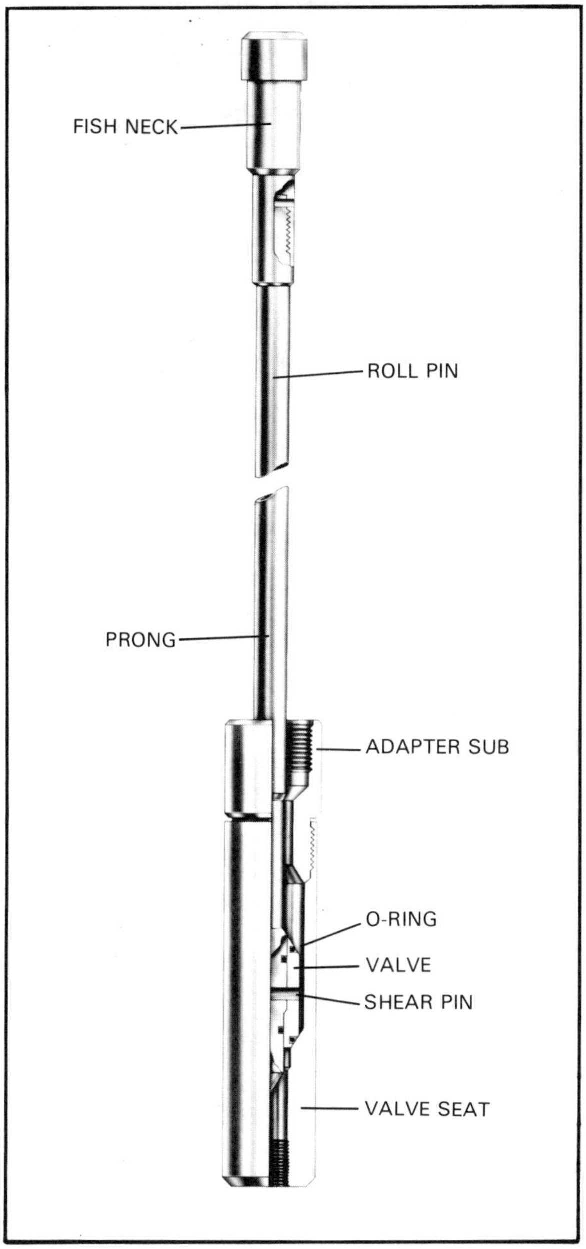

Figure 58. Plug choke (Courtesy of Otis Engineering)

Self-Test

5: Completion and Production Maintenance Operations

(Multiply each correct answer by five to arrive at your percentage of competency.)

Fill in the Blank

1. List three completion methods.

   (1) ______________________________

   (2) ______________________________

   (3) ______________________________

2. In a ______________________ completion, a string of casing is run and cemented across the productive zone.

3. List the three common methods of perforating.

   (1) ______________________________

   (2) ______________________________

   (3) ______________________________

4. The charges in a perforating gun are positioned in a ______________________ ______________ around the gun so that holes can be punched in all directions along the desired formation.

5. __________ ______________________ is the most commonly used perforating method.

6. ______________ ______________________ may not be used in a well with a temperature higher than 250°F.

True or False

7. _____ Mechanical perforating is used to perforate tubing but not casing.

8. _____ Swabbing is used to clean paraffin and other accumulations out of the wellbore.

9. _____ The most common completion is a temporary tubing string installed to the top of the producing zone.

10. ______ A landing nipple provides a seat at a known depth for various types of retrievable flow control equipment.

11. ______ A nonselective or no-go nipple is usually placed at or near the bottom of the tubing string.

12. ______ Downhole wireline equipment is used only on completions using a packer.

13. ______ The sliding sleeve valve is used to establish communication of fluids and pressure between the tubing and the tubing-casing annulus.

14. ______ Side pocket mandrels may be used as receivers for dummy valves, circulating valves, and chemical injection valves.

15. ______ A pulling tool is used to retrieve a mandrel control device assembly.

16. ______ Flow regulators prevent surface controls from freezing up when pressure increases at the surface.

Glossary

A

alligator grab *n:* a fishing device used to pick up relatively small objects like wrenches that have fallen or been dropped into the wellbore. The alligator grab has jaws similar to those of an alligator, which are pinned open before the tool is run into the well. The jaws snap shut over the fish when contact is made.

Amerada bomb *n:* a wireline instrument for measuring bottomhole temperature or pressure. It contains a clock-driven recording section and either a pressure element or a temperature element. It is thin enough to pass downhole through small tubing.

B

barefoot completion *n:* also called open-hole completion. See *open-hole completion.*

bottom wiper plug *n:* a device placed in the cementing head and run down the casing in front of cement to clean the mud off the walls of the casing and to prevent contamination between the mud and the cement.

bottomhole choke *n:* a device with a restricted opening placed in the lower end of the tubing to control the rate of flow. See *choke.*

bottomhole flow regulators *n:* See *bottomhole choke.*

Bourdon tube *n:* a flattened metal tube bent in a curve, which tends to staighten when pressure is applied internally. By the movements of an indicator over a circular scale, a Bourdon tube indicates the pressure applied.

bullet perforator *n:* a tubular device that, when lowered to a selected depth within a well, fires bullets through the casing to provide holes through which the formation fluids may enter the wellbore.

C

caliper log *n:* a record whereby the diameter of the wellbore is ascertained, indicating undue enlargement due to caving in, washout, or other causes. The caliper log also reveals corrosion, scaling, or pitting inside tubular goods.

chemical cutoff *n:* a method of severing steel pipe in a well by applying high-pressure jets of a very corrosive substance against the wall of the pipe. The resulting cut is very smooth.

choke *n:* a device with an orifice installed in a line to restrict the flow of fluids. Surface chokes are part of the Christmas tree on a well and contain a choke nipple, or bean, with a small-diameter bore that serves to restrict the flow. Chokes are also used to control the rate of flow of the drilling mud out of the hole when the well is closed in with the blowout preventer and a kick is being circulated out of the hole. See *adjustable choke, bottomhole choke,* and *positive choke.*

Christmas tree *n:* the control valves, pressure gauges, and chokes assembled at the top of a well to control the flow of oil and gas after the well has been drilled and completed.

conductor line *n:* a small-diameter conductive line used in electric wireline operations, such as electric well logging and perforating, in which the transmission of electrical current is required. Compare *wireline.*

D

deviation and azimuth indicator *n:* See *double recorder.*

deviation survey *n:* an operation made to determine the angle from which a bit has deviated from the vertical during drilling. There are two basic deviation-survey, or drift-survey, instruments: one reveals the angle of deviation only; the other indicates both the angle and the direction of deviation.

directional survey *n:* a logging method that records hole drift, or deviation from the vertical, and direction of the drift. A single-shot directional-survey instrument makes a single photograph of a compass reading of the drift direction and the number of degrees the hole is off vertical. A multishot-survey instrument obtains numerous readings in the hole as the device is pulled out of the well. See *directional drilling.*

directional survey instrument *n:* a tool that, when lowered into the wellbore, makes a photographic record of the angle and drift of the wellbore; that is, it records the number of degrees the hole is off vertical and the direction in which it is off vertical. Several types of instruments are available; some are capable of photographing only a single record – single shot survey instruments, while others are capable of making several records in one run – multishot survey instruments.

double recorder *n:* a device, containing an angle indicator, a timing element, and a replaceable chart, that is used to measure the direction and the degree of deviation as a hole is being drilled. The timing element goes off when the tool has come to rest just above the drill bit and descends on the sharply pointed angle indicator, which then punches a hole in the chart. The chart rotates 180°, and a second hole is punched to verify a correct reading of the resting angle of the tool.

F

fishing neck *n:* a tool used on a conductive wireline string so that the wireline will pull out of the fishing neck if the tool string becomes stuck. The fishing neck may also be used so that a fishing tool can be latched around it in order to retrieve a tool string that is stuck.

flow coupling *n:*a device placed in a tubing string to protect a landing nipple in the tubing string from the turbulence that can result from the placement of flow control devices in the landing nipples. Flow couplings are usually about two feet long and are made of high-grade alloy steel; their internal diameter equals that of the tubing.

free-point indicator *n:* a tool designed to measure the amount of stretch in a string of stuck pipe and to indicate the deepest point at which the pipe is free. The free-point indicator is lowered into the well on a conducting cable. Each end of a strain-gauge element is anchored to the pipe wall by friction springs or magnets, and, as increasing strain is put on the pipe, an accurate measurement of its stretch is transmitted to the surface. The stretch measurements indicate the depth at which the pipe is stuck.

G

gamma-ray log *n:* See *radioactivity well logging.*

gas-lift mandrel *n:* a device installed in the tubing string of a gas-lift well onto which or into which a gas-lift valve is fitted. There are two common types of mandrels. In the conventional gas-lift mandrel, the gas-lift valve is installed as the tubing is placed in the well. Thus, to replace or repair the valve, the tubing string must be pulled. In the sidepocket mandrel, however, the valve is installed and removed by wireline while the mandrel is still in the well, eliminating the need to pull the tubing to repair or replace the valve.

gin pole *n:* a pole (usually single) supported with guy wires and used with block and tackle to hoist equipment. On a drilling rig the gin pole is typically secured to the mast or derrick above the monkeyboard.

go-devil *n:* 1. a device that is inserted into a pipeline for the purpose of cleaning; a line scraper. Also called a pig. 2. a device that is lowered into the borehole of a well for various purposes such as enclosing surveying instruments, detonating instruments, and the like. *v:* to drop or pump a device down the borehole, usually through drill pipe or tubing.

H

hay pulley *n:* a pulley that is normally attached to the wellhead at a convenient place for the wireline to pass through as it comes from the stuffing box sheave before being spooled onto the wireline reel. The hay pulley prevents anẏ lateral forces from being exerted on the lubricator and the wellhead.

hydrostatic bailer *n:* a sand bailer that consists of a sealed cylinder with a rubber disk and a pierceing tool at the bottom. A sharp blow on the bottom ruptures the disk, allowing the hydrostatic pressure of the well fluid to force sand into the bailer.

I

impression block *n:* a block with lead or another relatively soft material on its bottom. It is made up on drill pipe or tubing at the surface, run into a well, and allowed to rest on a tool or other object that has been lost in the well. When the block is retrieved, the size, shape, and position of the fish are obtained from the examination of the impression left in the lead, and an appropriate fishing tool may be selected.

J

jar *n:* a percussion tool operated mechanically or hydraulically to deliver a heavy hammer blow to objects in the borehole. Jars are used to free objects stuck in the hole or to loosen tubing or drill pipe that is hung up. Blows may be delivered downward or upward, the jar being controlled at the surface. *v:* to apply a heavy blow to the drill stem by use of a jar.

jet cutoff *n:* a procedure for severing pipe stuck in a well by detonating special shaped-charge explosives similar to those used in jet perforating. The explosive is lowered into the pipe to the desired depth and detonated. The force of the explosion makes radiating horizontal cuts around the pipe, and the severed portion of the pipe is retrieved.

jet-perforate *v:* to create a hole through the casing with a shaped charge of high explosives instead of a gun that fires projectiles. The loaded charges are lowered into the hole to the desired depth. Once detonated, the charges emit short, penetrating jets of high-velocity gases that cut holes in the casing and cement and some distance into the formation. Formation fluids then flow into the wellbore through these perforations. See *bullet perforator* and *gun-perforate.*

K

knuckle joint *n:* a deflection tool, placed above the drill bit in the drill stem, with a ball and socket arrangement that allows the tool to be deflected at an angle; used in directional drilling. A knuckle joint is useful in fishing operations because it allows the fishing tool to be deflected to the side of the hole where a fish may have come to rest.

L

landing nipple *n:* a device machined internally to receive the removable locking devices used to position, lock, and seal subsurface production controls in tubing. A landing nipple provides a seat at a known depth, into which various types of retrievable flow control equipment can be set.

liner completion *n:* a well completion in which a liner is used to obtain communication between the reservoir and the wellbore.

line speed indicator *n:* an instrument calibrated to show the speed in feet per minute at which wireline is raised or lowered in the hole. The line speed indicator may be driven mechanically by the measuring wheel and placed next to the depth odometer, or it may be driven by a motor.

lubricator *n:* a specially fabricated length of casing or tubing usually placed temporarily above a valve on top of the casinghead or tubing head; used to run swabbing or perforating tools into a producing well; provides a method for sealing off pressure and thus should be rated for highest anticipated pressure.

M

mechanical perforating *n:* a procedure used to punch holes into the tubing string, one at a time, to circulate or to kill a well. Hydraulic jarring action fires a powder charge with enough force to drive a small button through the tubing, but not through the casing.

mechanical wireline *n:* See *nonconductive wireline.*

N

nonconductive wireline *n:* wireline used for operations that do not require that electric signals be relayed from downhole to the surface or from the surface to a downhole device.

P

paraffin *n:* a hydrocarbon having the formula C_nH_{2n+2} (e.g., methane, CH_4; ethane, C_2H_6). Heavier paraffin hydrocarbons (i.e., $C^{18}H_{38}$ and heavier) form a waxlike substance that is called paraffin. These heavier paraffins often accumulate on the walls of tubing and other production equipment, restricting or stopping the flow of desirable lighter paraffins.

perforated completion *n:* a well completion in which the production casing or liner is punctured to allow passage between the wellbore and the producing formation. Perforations are usually made with bullet- or jet-perforating guns.

plug choke *n:* a device used to plug the tubing at some point along the string. A plug choke may be set to test for packer leakage, to stop production when wellhead equipment must be removed, to test for tubing corrosion or other damage, and to separate zones for production or stimulation treatments.

production *n:* 1. the phase of the petroleum industry that deals with bringing the well fluids to the surface and separating them and with storing, gauging, and otherwise preparing the product for the pipeline. 2. the amount of oil or gas produced in a given period.

R

radioactivity well logging *n:* the recording of the natural or induced radioactive characteristics of subsurface formations. A radioactivity log, also known as a radiation log, normally consists of two recorded curves: a gamma ray curve and a neutron curve. Both help to determine the types of rocks in the formation and the types of fluids contained in the rocks. The two logs may be run simultaneously in conjunction with a collar locator in a cased or uncased hole.

resistivity well logging *n:* the recording of the resistance of formation water to natural or induced electrical current. The mineral content of subsurface water allows it to conduct electricity. Rock, oil, and gas are poor conductors. Resistivity measurements can be correlated to formation lithology, porosity, permeability, and saturation and are very useful in formation evaluation. See *electric well log.*

riser assembly *n:* See *lubricator.*

S

sample bailer *n:* a tool with a sharp, pointed end designed to slice into sand and a hollow cylinder to trap sand samples. The sample bailer is slammed into the bottom of the wellbore with enough impact to force sand into the bailer.

sand bailer *n:* a device used to remove sand from the wellbore. The bailer sucks up sand by creating a partial vacuum and traps the sand by closing at the bottom.

sand line *n:* a wireline used on drilling rigs and well-servicing rigs to operate a swab or bailer, to retrieve cores, or to run logging devices. It is usually $^9/_{16}$ of an inch (15 mm) in diameter and several thousand feet (metres) long.

self-potential curve *n:* also called spontaneous potential curve. See *spontaneous potential curve.*

side-pocket mandrel *n:* See *gas-lift mandrel.*

sidewall coring *n:* a coring technique in which core samples are obtained from a zone that has already been drilled. A hollow bullet is fired into the formation wall to capture the core and then retrieved on a flexible steel cable. Core samples of this type usually range from ¾ to $1^3/_{16}$ inches (20 to 30 mm) in diameter and from ¾ to 4 inches (20 to 100 mm) in length. This method is especially useful in soft-rock areas.

sidewinder unit *n:* a type of wireline unit in which power is transferred by means of a pulley and a belt from the left rear wheel of the wireline truck to the spool of wire that is pulled from the left side of the truck. The spool is used to run the wireline downhole and bring it back up again.

sinker bar *n:* a heavy weight or bar placed on or near a lightweight wireline tool. The bar provides weight so that the tool will lower properly into the well.

slick line *n:* also called wireline. See *wireline.*

sliding-sleeve nipple *n:* a special device placed in a string of tubing and operated by a wireline tool (1) to open or close orifices, thus permitting circulation between the tubing and annulus, or (2) to open or shut off production from alternate intervals in a well.

sonde *n:* a logging tool assembly, especially the device in the logging assembly that senses and transmits formation data.

spaghetti *n:* tubing or pipe with a very small diameter.

spontaneous potential curve *n:* a measurement of the electrical currents that occur in the wellbore when fluids of different salinities are in contact. The SP curve is usually recorded in holes drilled with freshwater-base drilling fluids. Also called self-potential curve.

stem *n:* See *sinker bar.* Also called a weight bar.

string shot *n:* an explosive device that uses primacord, a textile-covered fuse with a core of very high explosive, to create an explosive jar inside stuck pipe or tubing to back off the pipe at the joint immediately above the stuck pipe. See *shot.*

stroke jar *n:* a mechanical percussion tool that can deliver either upward or downward impact. See *jar.*

swab *n:* a hollow, rubber-faced cylinder mounted on a hollow mandrel with a pin joint on the upper end to connect to the swab line. A check valve that opens upward on the lower end provides a way to remove the fluid from the well when pressure is insufficient to support flow. *v:* to operate a swab on a wireline to bring well fluids to the surface when the well does not flow naturally. Swabbing is a temporary operation to determine whether or not the well can be made to flow. If the well does not flow after being swabbed, a pump is installed as a permanent lifting device to bring the oil to the surface.

swage *n:* a tool used to straighten damaged or collapsed casing in a well.

T

tensile strength *n:* the greatest longitudinal stress that a metal can bear without tearing apart. Tensile strength of a metal is greater than yield strength.

top wiper plug *n:* a device placed in the cementing head and run down the casing behind cement to clean the cement off the walls of the casing and to prevent contamination between the cement and the displacement fluid.

tubing *n:* small-diameter pipe that is run into a well to serve as a conduit for the passage of oil and gas to the surface.

tubing broach *n:* a tool with three consistently larger spools used to gradually enlarge tubing diameter. The tubing broach may be used to either enlarge the diameter of obstructed tubing or to cut paraffin or other substances that may be inside the tubing.

tubing end locator *n:* a tool containing a spring-loaded finger, or *dog,* that pops out when the tool reaches the bottom end of the tubing string. The extended dog prevents the tool from being raised back into the tubing, thus indicating to an operator on the surface that the bottom of the tubing has been reached. The tool is removed from the tubing by jarring upward with enough force to shear a brass pin and allow the dog to rotate downward and into a slot in the body of the tubing end locator.

tubing gauge *n:* a tool designed to cut away scale, calcium, small deposits of paraffin, and other debris in an obstructed length of tubing. The tool is also used to measure, or gauge, the internal diameter of tubing.

W

weight bar *n:* See *sinker bar*. Also called stem.

well completion *n:* the activities and methods necessary to prepare a well for the production of oil and gas; the method by which a flow line for hydrocarbons is established between the reservoir and the surface. The method of well completion used by the operator depends on the individual characteristics of the producing formation or formations. Such techniques include open-hole completions, sand-exclusion completions, tubingless completions, and miniaturized completions.

wellhead *n:* the equipment installed at the surface of the wellbore. A wellhead includes such equipment as the casinghead and tubing head. *adj:* pertaining to the wellhead (e.g., wellhead pressure).

well logging *n:* the recording of information about subsurface geologic formations. Logging includes records kept by the driller and records of mud and cutting analyses, core analysis, drill stem tests, and electric, acoustic, and radioactivity procedures. See *driller's log, mud analysis, core analysis, drill stem test, electric well log, acoustic log,* and *radioactivity log.*

wireline *n:* a small-diameter metal line used in wireline operations; also called slick line. Compare *conductor line.*

wireline blowout preventer *n:* a device containing a valve that closes around the wireline to shut in the well if high pressure threatens a blowout. Also called a wireline valve.

wireline grab *n:* a tool used to fish broken wireline out of tubing. Two or three barbed prongs extend along the inside of the U-shaped housing. Wireline becomes caught on these prongs so that it can be retrieved.

wireline operations *n pl:* the lowering of mechanical tools, such as valves and fishing tools, into the well for various purposes. Electric wireline operations, such as electric well logging and perforating, involve the use of conductor line, which in the oil patch is commonly but erroneously called wireline.

wireline socket *n:* the connection used to attach wireline to the tool string. The wireline socket contains a rotating disk to allow the tool string to swivel horizontally and a small spring to cushion the line and to prevent the line from breaking during heavy operations.

wireline spear *n:* a special fishing tool fitted with prongs to catch and recover wireline that has broken off and been left in a well.

wireline tool string *n:* the string of tools and equipment that is attached to conductive or nonconductive wireline and lowered downhole to perform wireline operations.

wireline wiper *n:* a flexible rubber wiper used to scrape mud or oil from a wireline as it is pulled from a hole.

wire rope *n:* a cable composed of steel wires twisted around a central core of fiber or steel wire to create a rope of great strength and considerable flexibility. Wire rope is used as drilling line (in rotary and cable-tool rigs), coring line, servicing line, winch line, and so on. It is often called cable or wireline; however, wireline is a single, slender metal rod, usually very flexible. Compare *wireline.*

Self-Test Answer Key

1: Development of Wireline Services

1. Wireline
2. wireline operations
3. cementing
4. clothesline
5. conductive
6. (1) Slick
   (2) Braided
   (3) Corrosion-resistant
7. Tensile strength
8. (1) With its lower bend radius, a small-diameter line can be wound around a smaller reel drum, increasing convenience and decreasing cost.
   (2) Small-diameter line is lighter and requires less braking capability on the surface engine that powers the takeup reel.
   (3) Small-diameter wire presents a small cross-sectional area when an operation is performed under high pressure. Less weight is necessary to move it through pressure containment devices.
9. F
10. T
11. F
12. T
13. T
14. F
15. T

2: Wireline Equipment

1. (1) The depth to which the line is to be run
   (2) Downhole pressure
   (3) Downhole temperature
   (4) Casing diameter and weight
2. (1) A reel of slick or conductor line
   (2) A power system enabling operators to raise and lower the line and to control its speed
   (3) Instruments indicating weight and depth of the downhole tools
   (4) Pressure-containment equipment
   (5) A tool string to provide a way to attach tools to the line
3. manual
4. Advantage: It allows the same wide range of power and speed control as the engine of a truck.
   Disadvantage: It requires two operators – the driver and an outside operator – to keep track of the line and its progress.
5. (1) Hydraulic
   (2) Electric
6. (1) Weight
   (2) Depth
   (3) Line speed
7. (1) Wireline socket
   (2) Stem
   (3) Weight bar
   (4) Jar
   (5) Knuckle joint
   (6) Wireline tool
8. (1) Fishing neck
   (2) Head
   (3) Turn buckle
9. (1) Wellhead pressures are high
   (2) Dangerous hydrogen sulfide gas is present in the well
10. blowout preventer
11. (1) Ginpole
    (2) Block and tackel
    (3) Upper master valve
    (4) Wing valve
    (5) Wireline valve
    (6) Lubricator
    (7) Stuffing box
    (8) Hay pulley
12. (1) Wash the line as soon as it comes out of the hole
    (2) Oil the line before it is stored
    (3) Transfer the line correctly from the spool to the reel
    (4) Place the lowest possible weight on the line
13. T
14. F
15. F
16. T
17. T
18. F
19. F
20. T
21. T

3. Diagnostic Operations

1. drill bit; workover; artificial lift; casing
2. to run a recording instrument through the drill pipe on standard measuring-sized slick wireline after every few hundred feet have been drilled.
3. (1) Angle indicator
   (2) Timing element
   (3) Replaceable chart
4. sonde
5. (1) Electrical
   (2) Radioactive
   (3) Acoustic
6. (1) Self-potential (SP)
   (2) Caliper
   (3) Gamma ray
   (4) Resistivity
   (5) Porosity
7. (1) Logging
   (2) Drill stem testing
   (3) Mud sampling
8. Any three of the following:
   (1) A distortion or obstruction in the tubing string or casing may block the flow.
   (2) Sand can enter the wellbore and can block the upward flow of oil or gas.
   (3) A tubing hole, caused by corrosion or pitting, may deflect the hydrocarbons out of the tubing and into the tubing-casing annulus.
   (4) A length of pipe may cave in, or collapse, because of extreme force from the adjacent formation.
9. tubing end locator; sand bailer

4: Troubleshooting Operations

1. (1) Freeing a length of tubing, casing, or drill pipe just above the point where it is stuck downhole so that the stuck portion can then be retrieved;
   (2) Repairing or conditioning tubing;
   (3) Removing sand or paraffin deposits from strategic downhole locations;
   (4) Fishing for dropped tools or various pieces of junk downhole.
2. (1) String-shot backoff
   (2) Jet cutter
   (3) Chemical cutter
3. electrical impulses
4. jet cutter
5. chemical cutter
6. swaging
7. tubing gauge
8. sand bailer
9. hydrostatic sand bailer
10. perforations
11. (1) Impression block
    (2) Wireline grab
    (3) Overshot
    (4) Magnetic fishing tool
    (5) Alligator grab

5: Completion and Production Maintenance Operations

1. (1) Barefoot completion
   (2) Liner completion
   (3) Perforated completion
2. perforated
3. (1) Jet perforating
   (2) Bullet perforating
   (3) Mechanical perforating
4. vertical spiral
5. Jet perforating
6. Bullet perforating
7. T
8. F
9. F
10. T
11. T
12. F
13. T
14. T
15. T
16. F